T0185624

Association for Women in Mathematics Series

Volume 23

Series Editor
Kristin Lauter
Microsoft Research
Redmond, Washington, USA

Association for Women in Mathematics Series

Focusing on the groundbreaking work of women in mathematics past, present, and future, Springer's Association for Women in Mathematics Series presents the latest research and proceedings of conferences worldwide organized by the Association for Women in Mathematics (AWM). All works are peer-reviewed to meet the highest standards of scientific literature, while presenting topics at the cutting edge of pure and applied mathematics, as well as in the areas of mathematical education and history. Since its inception in 1971, The Association for Women in Mathematics has been a non-profit organization designed to help encourage women and girls to study and pursue active careers in mathematics and the mathematical sciences and to promote equal opportunity and equal treatment of women and girls in the mathematical sciences. Currently, the organization represents more than 3000 members and 200 institutions constituting a broad spectrum of the mathematical community in the United States and around the world.

Titles from this series are indexed by Scopus.

More information about this series at http://www.springer.com/series/13764

Mary Lee • Aisha Najera Chesler

Editors

Research in Mathematics and Public Policy

 Springer

ASSOCIATION FOR
WOMEN IN MATHEMATICS

Editors
Mary Lee
RAND Corporation
Santa Monica, CA, USA

Aisha Najera Chesler
RAND Corporation
Santa Monica, CA, USA

ISSN 2364-5733   ISSN 2364-5741   (electronic)
Association for Women in Mathematics Series
ISBN 978-3-030-58750-5   ISBN 978-3-030-58748-2   (eBook)
https://doi.org/10.1007/978-3-030-58748-2

This Springer imprint is published by the registered company Springer Nature Switzerland AG
The registered company address is: Gewerbestrasse 11, 6330 Cham, Switzerland

Women in Mathematics and Public Policy Workshop participants, with keynote speakers Dr. Lucy Jones and Dr. Kristin Lauter

Debra Knopman speaks to workshop participants at networking dinner

Preface

The articles in this volume are proceedings papers inspired by a Workshop for Women in Mathematics and Public Policy, held on January 22–25, 2019 at the Institute for Pure and Applied Mathematics and the Luskin Center at the University of California, Los Angeles. The workshop was created to promote and develop women at all levels of their careers as researchers in mathematics and public policy. The idea was modeled after other successful Research Collaboration Conferences for Women, where junior and senior women come together at week-long conferences held at mathematics institutes to work on pre-defined research projects. Our workshop focused on how mathematics can be used in public policy research and was designed to foster collaborative networks for women to help address the gender gap in mathematics and science.

As mathematicians who work at a nonprofit research institution that helps improve policy and decision-making through objective analysis, we are well aware of the effect that public policy can have on important aspects of our lives, including health, education, security, and infrastructure. We also understand the need for research opportunities that provide support and promote networking to women in math, science, and policy. It seemed natural to combine these in a collaborative workshop to explore public policy questions that could benefit from mathematical thinking. Though the number of policy research questions that can be informed by advanced mathematical techniques is vast, the workshop focused on two impactful themes: cybersecurity and climate change. Both of these transdisciplinary areas of public policy research benefit from the expanded application of mathematical disciplines such as mathematical modeling, machine learning, and statistics. To the best of our knowledge, this was the first workshop at the intersection of math and public policy developed by and for women. We hope that these types of events and proceedings will continue to showcase the depth of skill of women in mathematics and policy at all career levels.

The workshop was sponsored by the Association for Women in Mathematics, the National Science Foundation, the Luskin Center and the Institute for Pure and Applied Mathematics at UCLA, and the RAND Corporation. We would like to thank our sponsors, workshop participants, project leads, co-organizers, and reviewers, without whom this research would not have been done. We are grateful to Daniel Apon, Amanda Back, Laura Baldwin, David Catt, Anita Chandra, Cynthia Dion-Schwarz, Maria D'Orsogna, Michael Hansen, Ted Harshberger, Lucy Jones, Lily Khadjavi, Debra Knopman, Kristin Lauter, Magnhild Lien, Gretchen Matthews, Caolionn O'Connell, Stacy Orozco, Osonde Osoba, Christopher Pernin, Ami Radunskaya, Christian Ratsch, Allison Reilly, Melissa Rowe, Dimitri Shlyakhtenko, Abbie Tingstad, Luminita Vese, and Henry Willis for their expertise and support.

Santa Monica, CA, USA   Mary Lee
                       Aisha Najera Chesler

Contents

Exploring the Application of Machine Learning for Downscaling Climate Projections

Kristin Van Abel, Amanda Back, M. Kathleen Brennan,
Oriana S. Chegwidden, Mimi Hughes, Marielle Pinheiro, and Cecilia M. Bitz

Abstract Policy makers need information about future climate change on spatial scales much finer than is available from typical climate model grids. New and creative methods are being advanced to downscale climate change projections with statistical methods. Important requirements are to reliably downscale the climate parameter means, variability, extremes and trends, while preserving spatial and temporal correlations and permitting uncertainty quantification. In this proof-of-concept study, datasets derived from both observations and climate models were used together to train and test statistical methods. Two machine learning techniques—artificial neural networks and random forests—were tested on the problem of using coarse-scale climate projections (here represented by ERA-Interim reanalyses) to create temperature predictions at specific locations in areas of complex terrain. The methods are trained on and validated by temperature readings from mesonet weather stations in Colorado. This work has implications for fire prevention and water resources management, among other applications.

K. Van Abel (✉)
RAND Corporation, Santa Monica, CA, USA
e-mail: kvanabel@rand.org

A. Back
Cooperative Institute for Research in the Atmosphere, NOAA/OAR/ESRL/Global Systems Division, Colorado State University, Fort Collins, CO, USA
e-mail: amanda.back@colostate.edu

M. K. Brennan · O. S. Chegwidden · C. M. Bitz
University of Washington, Seattle, Washington, USA
e-mail: mkb22@uw.edu; orianac@uw.edu; bitz@atmos.washington.edu

M. Hughes
NOAA Earth Sciences Research Laboratory, Physical Sciences Division, Boulder, CO, USA
e-mail: mimi.hughes@noaa.gov

M. Pinheiro
University of California, Davis, Davis, CA, USA
e-mail: Mcpinheiro@ucdavis.edu

© The Author(s) and the Association for Women in Mathematics 2020
M. Lee, A. Najera Chesler (eds.), *Research in Mathematics and Public Policy*, Association for Women in Mathematics Series 23, https://doi.org/10.1007/978-3-030-58748-2_1

1

1   Introduction

Understanding a changing climate at scales relevant to human activity and natural ecosystems is complicated by the heterogeneity of terrain, meteorology, and hydrology. Typically, climate studies make use of global climate models (GCMs) with spatial resolutions of 100–200 km. However, this resolution is too coarse to accurately represent the climate in areas of significant topographic variability or coastal areas, as well as certain meteorological phenomena whose dynamics operate on finer scales such as mesoscale convective systems and tropical cyclones (e.g. [1, 2]).

Local and regional policy decisions for a wide range of applications require higher resolution climate information than what is available in GCMs, which has led to a demand for regional precipitation, snow, and temperature data at finer scales. Annual totals of precipitation, which are affected by the occurrence (or lack) of a few major events, can vary substantially on a year-to-year basis, and future projections depend on the region, with both increases and decreases seen over the extent of the globe [3]. In addition to the variability and uncertainties inherent in climate dynamics, there can be large differences among extreme precipitation totals projected in the various GCMs, reflecting different representations of climate dynamics and the inevitable errors arising from any type of numerical model of a complex system [4]. Some of the spread in model output is due to the parameterization of sub-grid processes; native coarsely-resolved GCM output cannot account for the effects of topography on temperature and precipitation in a given watershed.

Even with improved GCM spatial resolution, these multiple sources of uncertainty mean that hyper-localized data are difficult to appropriately project on a time and spatial scale that is relevant and useful to policymakers. Water managers need finer scale information on these variables for projections of future water resource availability (e.g., [5]) and management of stormwater [6]. Projections of wildfire regimes also require more spatially and temporally resolved precipitation data (in addition to wind) at smaller scales, especially in regions of complex terrain [7]. Similar needs are present in other fields such as agricultural, biological, ecosystem services, and urban planning sectors.

The technique that translates coarse-scale GCM climate information to a finer spatial scale for relevance to local applications is termed "downscaling." Climate scientists use two types of downscaling techniques: dynamical and statistical. In brief, dynamical downscaling relies on Regional Climate Models (RCMs) to achieve finer spatial scales. RCMs are regional numerical weather prediction (NWP) models forced by GCM output at the lateral boundaries to downscale GCM output. Thus, their representation of future climate at the regional scale depends on the physics of the NWP model, the large-scale conditions from the GCM, and their own high-resolution representation of the terrain. RCMs have the advantage of using NWP physics to constrain the relationships between large and regional scales but are limited by the fidelity of the RCM's physics and the computational cost associated with high-resolution NWP. In contrast to the RCM's mathematical representation of physical processes, statistical downscaling uses historical observational data to empirically build relationships between large and small scales, capturing complex terrain features. Simple statistical downscaling approaches build the statistical relationships between the variable of interest at large scale and local scale (i.e., go from coarse-scale precipitation to fine-scale precipitation, e.g., [8]), but multivariate techniques are also possible though less widely used because of their complexity.

Improved downscaling techniques of either type could offer more credible and constrained climate projections to decision makers. The literature shows that the use of multiple downscaling techniques within one downscaling scheme will likely widen the spread of potential climate futures provided to decision-makers [7, 9, 10]. However, this approach also provides a better view of the range of uncertainty against which decision makers must plan. By improving downscaling techniques, decision makers can build confidence in the range of potential futures to consider in planning efforts. Further, by building more computationally-efficient downscaling techniques, scientists can reduce the costs associated with producing a set of downscaled projections and broaden their accessibility to communities that lack the technical resources to use complex models to inform decisions on stormwater infrastructure sizing or water storage, for example.

One promising new technique for statistical downscaling is machine learning, an automated model-building technique that stems from artificial intelligence methods. It is rooted in the idea that computer algorithms can "learn" from data, are capable of pattern recognition, and can perform certain tasks or make decisions with little programming or human intervention. The iterative aspect of machine learning is important; exposure to new data allows the model to adapt without user intervention. The science behind machine learning is not new, but improvements in computing technologies have spurred the application of machine learning algorithms to new problems.

At a very high level, there are three kinds of machine learning algorithms: (1) supervised, which involves making predictions based on a set of examples (e.g., using historical stock prices to make predictions about the future), and in which the functional relationships (linear or logarithmic, for example) between predictors and predictand is not imposed, as in regression analysis, but determined during model training; (2) unsupervised, in which input data are uncategorized and machine learning is used to organize the data and provide an idea of the structure; and (3) reinforcement learning where an algorithm receives an "award signal" after making a decision that indicates how good the decision was (commonly applied in robotics) [11].

There are several considerations for choosing a machine learning method and algorithm. We outline a few considerations below (see [12, 13]):

- Performance/accuracy: What will the prediction or projection be used for?
- Comprehensibility: Who needs to be able to interpret the model?
- Cost/time constraints: How fast does the training of the algorithm need to be? How fast does the execution of the trained algorithm need to be?

- Complexity of the data/area of application: Do data come from disparate sources or change rapidly? What is the nature of the correlations and other relationships in the data?

Machine learning has been applied in several industries including financial services (e.g., risk analytics, fraud prevention), health care (e.g., improving diagnosis and treatment, identifying risk factors), retail (e.g., marketing campaigns and recommendations based on buying history), energy (e.g., predicting and finding new energy sources, predicting equipment failures), manufacturing (e.g., predicting maintenance and condition monitoring), transportation (e.g., improving efficiency of routes, predicting problems or future areas of congestion), and is gaining momentum within climate science [14, 15]. In this proof-of-concept study, the authors (who had not previously worked together) trained and tested machine learning algorithms to downscale climate data and compared the output to real temperature observations during a 4 day workshop. This study is an example of researchers from diverse technical backgrounds working collectively on a challenging problem: some of the project participants had never used Python and others had never used machine learning techniques.

The remainder of the paper is structured as follows: Section 2 describes the data and methods used. Section 3 presents the results of this pilot effort. Section 4 provides a discussion of the results.

2 Data and Methods

This study explored whether machine learning techniques could be used to improve downscaling of climate data. Specifically, given a climate projection on a coarse grid, this work investigates how accurately conditions at precise locations within the domain can be predicted. To aid in prediction, historical data—past coarsely-gridded reanalysis data[1] and observations at the specific locations where predictions will be made—are used to build statistical relationships between the large and small scales. Machine learning, which in this application can be regarded as a type of statistical downscaling, is used both to find relationships and to generate predictions.

---

[1]Reanalysis is a scientific method that combines observations and a numerical model of one or more aspects of an earth system to generate an estimate of the state of the system. Reanalysis products are used in climate research to represent how weather and climate are changing over time.

2.1 Description of Datasets Used

Surface temperature data was chosen as the focus of this proof-of-concept study because it has a strongly resolved signal of change in climate models. Other variables of interest include snow accumulation and total precipitation. Time did not allow for the prediction of these variables to be attempted during the workshop, but they remain relevant for consideration in future research.

For the coarsely-gridded climate model output, monthly European Centre for Medium-Range Weather Forecasts (ECMWF) Re-Analysis data (ERA-Interim) were used [16]. ERA-Interim (ERAi) data cover January 1979 through August 2019, and includes surface temperatures, among many variables, on a global grid of about 79 km horizontal spacing. As a reanalysis product, the ERAi temperature is a combination of model output and observational data: the model fields form a background into which observations are assimilated to produce the best possible estimate of climate conditions at a past time. Because observational data are available earlier than reanalyses,[2] this product is not a projection, but instead represents a baseline from which to experiment with genuine projections. The research team had hoped to replace ERAi with Community Earth System Model Large Ensemble projections (CESM-LE; [17]) in a subsequent round of experiments but did not have time during the workshop. This remains an avenue for future work.

For the surface temperature observations, mesoscale network (mesonet) data were used. These data are collected by surface weather stations, including at airports and academic institutions, at hourly or higher cadences. The data collected by these stations vary in quality for many reasons, including the use of different sensing equipment, maintenance routines, and siting [18]. NOAA (https://madis.ncep.noaa.gov, https://www.ncdc.noaa.gov) provides mesonet data augmented with quality control flags, but these data sets were unavailable during the workshop.[3] Instead, we identified data archived by Iowa State University (https://mesonet.agron.iastate.edu) as suitable for our purposes. Some quality control issues, such as a consistent bias, should not impede the ability of machine learning algorithms to successfully detect patterns and relationships. However, the lack of quality control on these data sets may still impact results, and future studies should endeavor to use quality-controlled data when possible. To align the observation data with the monthly mean data from ERAi, each station's temperature data were averaged by month, yielding one "observation" per station per month.

---

[2]ERAi products are available two or more months later to allow for thorough collection of observations and for quality control.

[3]The workshop occurred in January 2019 during a U.S. government lapse in appropriations.

2.2 Machine Learning Techniques

Two machine learning techniques, Random Forests (RF) and Artificial Neural Networks (ANN), are the focus of this study. The brief overview of these methods in this subsection follows the far more extensive treatment by Géron [19].

The RF method works by training many decision trees simultaneously, each on a random subset of inputs, also referred to as features. In the context of this study, a decision tree is a branching sequence of binary evaluations of the relationship between the set of predictors (i.e., reanalysis surface temperature data) and predictand (i.e., downscaled monthly mean temperature data). When new data are input to the trained model, all trees are applied to the input data, and the results from all trees are averaged to yield a single output.

ANNs are designed to mimic the way networks of neurons function: each node, or artificial neuron, performs a simple task while the network as a whole is exponentially more complex. In the ANN framework, each node performs a single mathematical operation (such as a step function or a logarithm); the nodes are arranged in layers and nodes in different layers are linked by weighted connections, analogous to synapses, whose weights are determined during training. Input data passes through each layer according to the synaptic weights, with each node performing its operation on the weighted sum of incoming data and then transmitting a single value, to produce a single numerical value as output. ANNs differ in how they are trained; this study implemented a multilayer perceptron (MLP), which uses backpropagation to convey error information through the network.

A substantial body of research exists on the subject of downscaling using machine learning methods. We highlight a few works relevant to this study. ANN has been used in the past to downscale daily average temperatures [20–22], daily temperature extremes [23, 24], and monthly mean maximum and minimum temperatures [25]. RF has been used to downscale daily average temperatures [20] and daily low temperatures [26].

Reanalysis products are commonly used in these studies for the large-scale climatology during model calibration and validation (see [20, 22, 23, 26]). Some studies then apply the most successful downscaling methods to downscale future climate scenarios from GCMs (see [21, 24, 25, 27]).

In several studies a large suite of predictor variables from the climatology are considered, though Goyal and Ojha [25] used only three fields (air temperature, zonal winds, and meridional winds at 925 hPa) to downscale monthly mean maximum and minimum temperatures, and Weichert and Bürger [22] used only the geopotential height of the 500 hPa layer and air temperature at 850 hPa to downscale daily average temperatures. Holden et al. [26] achieved temperature downscaling at locations distinct from those used for training by adding geographical features (altitude and an index of local terrain complexity) to the set of predictors along with reanalysis variables.

To reduce the high dimensionality of predictor sets, several studies used principal component analysis [22, 24–26], though Pang et al. [20] found that random forests did not need the dimensionality reduced. Some studies trained separate models on individual seasons [21, 26, 27]. It is common to use the most recent quarter [15, 23] or third [20] of the available data for validation.

In contrast to past studies, this research effort implements RF and ANN to downscale monthly mean temperature from reanalysis data using only reanalysis surface temperature, projected onto principal components, as a predictor. The RF algorithm was chosen because the underlying decision trees are easy to interpret, data do not need to be normalized (although this study did normalize the data), and RF can handle missing data since each individual tree depends on only a few features; when data are missing, trees that rely on that data can be omitted from the calculation. Note that we were not missing data for this study.

The ANN algorithm was chosen because it can handle complex, nonlinear relationships, can find unknown functional relationships between variables (e.g., can generalize well), and does not impose restrictions on input data.

Linear regression (LR), a less sophisticated form of ML, was also implemented for comparison purposes. While past studies have also compared ML methods to more standard linear approaches [21–25], this study's direct comparison of ANN to RF has few precedents (see [20]). In using surface temperature reanalysis data only, this study's RF experiment is similar to Crawford et al. [15], who used a random forest to blend temperature projections from an ensemble of GCMs, achieving higher accuracy than did an arithmetic mean of the ensembles. This research effort also experimented with using a single trained model for all months, with the annual cycle removed, and with training a model for the October through December months only.

Python was used in this study because several Python-based machine learning software packages already exist and have been extensively tested. Analysis was performed using Jupyter notebooks [28] and the code was tracked using Git to support version control. All of the code from this effort is available at https://github. com/orianac/women_publicpolicy_downscaling for use by anyone in the research community.

2.3 Data Preparation

From the global ERAi climate data, a 30x40 grid was chosen encompassing a large swath of the intermountain West (see Fig. 1) to serve as the coarse domain. Within this domain, the research team trained and tested on 15 weather stations in Colorado, selected for its complex terrain. Although the weather stations are near the edge of the chosen ERAi domain, shifting the ERAi domain to be centered on Colorado did not qualitatively affect outcomes.

The algorithms described in Sect. 2.2 require input data describing the large-scale climate projection. The 1200 grid points in the ERAi domain represent a significant computational burden for the RF and ANN algorithms; instead of using data at all points as features, the ERAi data were instead projected onto principal components

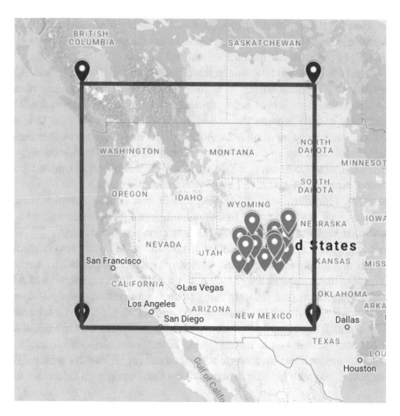

Fig. 1 Locations of station data used within coarse reanalysis domain. Red markers and box: corners and edges of large scale ERAi domain used for climate reanalysis data; Blue markers: weather stations in Colorado providing observational data. Source: Google Maps

(PCs), of which the first 20 were used to construct truncated series approximating the full climate reanalyses.

To construct the principal components, data from the period October 1979 through September 2005 were used, with later years held back for testing. First, monthly averages were computed over all training years and the entire domain, yielding a single average temperature value for the domain for October, one for November, and so on. Next the corresponding monthly mean was subtracted from each month's temperature field, a function of latitude and longitude, recasting the ERAi temperature data in terms of anomalies from the monthly mean. Finally, these gridded anomalies were used to compute PCs. This set of PCs represent the highest modes of variability across all months of the training time period. Each monthly temperature anomaly field T can then be approximated by a truncated series of PCs P_k and coefficients θ_k,

$$T(x, y, t) \approx \theta_0(t) + \theta_1(t)P_1(x, y) + \theta_2(t)P_2(x, y) + \ldots, \tag{1}$$

where x and y represent longitude and latitude, t represents the month, and θ_0 represents the offset of the mean temperature for month t from the monthly mean for all training years. From the series representation of the temperature anomaly field, only the coefficients θ_k are provided to the ML algorithms as features; the temperature fields, temperature anomalies, and PCs are never directly analyzed by the ML implementations. Thus the high dimensionality of a grid of 1200 data points is reduced to a set of at most 21 coefficients.

In addition to averaging the mesonet observational data into monthly reports, monthly means were subtracted from the observations to create observed anomalies consistent with the ERAi anomalies.

3 Results

The ML algorithms described in Sect. 2.2 were tested in three different experiments. In the first experiment, all data from October 1979 through September 2017 were used to train the ML models, and then the features from October 2004 through September 2017 were reused as inputs for testing. Thus, in this experiment the ML models were fed the features and observations for the test period, and also data from the non-test period. This is an idealized test case because in practical applications the correct answers are not available; the ability of the algorithms to produce new, accurate output is not measured by this experiment. The question in this case was whether the models would, given a set of features, identify that input with the data already analyzed and produce the answer the models were given in training.

In the second experiment the testing and training data sets are distinct, where data from October 1979 through September 2004 was used to train the models, and data from October 2004 through September 2017 was used to test the models. In this experiment the observation data from mesonet stations is replaced with ERAi data at a point in the domain.

Let an ERAi grid point be given by (x_i, y_j); then an ERAi temperature reanalysis anomaly at that point is approximated, as in Eq. (1), by,

$$T\left(x_i, y_j, t\right) \approx \theta_0(t) + \theta_1(t) P_1\left(x_i, y_j\right) + \theta_2(t) P_2\left(x_i, y_j\right) + \ldots \qquad (2)$$

Because the PCs are time-independent, the value of a PC at a specific point is a constant scalar. Furthermore, the ERAi temperature anomaly at that point is a function of time only. Equation (2) can be rewritten,

$$T_{ij}(t) \approx \theta_0(t) + \theta_1(t) P_{1ij} + \theta_2(t) P_{2ij} + \ldots, \qquad (3)$$

to emphasize that the kth PC at the point (x_i, y_j) is a constant scalar value and the temperature anomalies at the point (x_i, y_j) form a time series. In the second experiment, the set of coefficients $\{\theta_k\}$ at each time constitute the features provided to the model, while T_{ij} are the desired outputs. Let $n + 1$ be the number of features that will be input (up to 21 for our experiments), then the training data set for the second experiment can be expressed by $\{\theta_k(t),\ k = 0,\ldots,n;\ T_{ij}(t);\ t \leq$ September 2004}; the testing dataset is $\{\theta_k(t),\ k = 1,\ldots,n,\ t >$ September 2004}; and model outputs during testing are compared to $\{T_{ij}(t),\ t >$ September 2004}. The ML models can perform extremely well in this experiment if they learn $\{P_{kij}\}$ during training, but they are not configured to identify this specific pattern.

The third experiment is the most realistic of the three. In this experiment, the training data covers the months October, November and December from 1979 through 2004, and the testing data covers the months October, November and December from 2005 through 2016. Mesoscale observations are used as outputs—both provided to the models for the training time period and compared to model output for testing. This configuration most closely resembles real-world downscaling applications in which past climate model output and weather station data are known, as are future (coarse-resolution) climate model projections, and weather station data at some future time are to be estimated.

3.1 Experiment 1: Predicting Training Data

As an initial test of both models, RF and ANN algorithms were trained on all available temperature data (October 1979 to September 2017) to predict temperature for October 2004 to September 2017. Thus, the testing data were a subset of the training data. This is an idealized test case in which the models should perform well and thus the results indicate the upper bound of performance of these models for downscaling when observation data are unavailable.

Figure 2 shows a time series of station observation data for the testing time period, the RF-predicted temperature anomaly, and the ANN-predicted temperature anomaly at one Colorado location (U.S. Air Force Academy weather station) throughout the testing period. Note that the RF curve lies beneath the observation curve. Figure 3 shows the same data in a scatter plot of predicted versus observed temperature anomaly for the same station and time period. The RF-predicted temperature falls on a tight one-to-one line with a slight high bias relative to observations. The ANN also predicts temperature well, albeit with greater error than the RF. Here it is clear that the RF predictions out-perform those from the ANN on the training data. However, the excellent fit of the RF predictions suggests that the RF model may be over-fitting and may not perform well on data outside the training data set. This could be due to the number of degrees of freedom the RF model is configured to have—the number may be too large relative to the amount of data used for training [19].

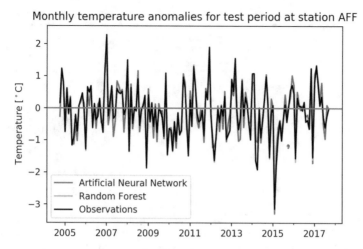

Fig. 2 RF and ANN time series

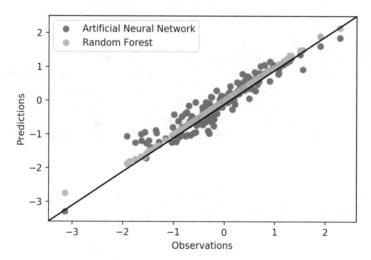

Fig. 3 Comparison of artificial neural network and random forest fit

3.2 Experiment 2: Predicting ERAi Data

Next, the performance of the models were tested for predicting the ERAi temperature anomaly at a grid point from the large-scale reanalysis. In this experiment, the training data set consisted of the temperature anomaly of the ERAi reanalysis at the chosen grid point, along with the coefficients θ_k, for the training time period of October 1979 to September 2004. For the testing data (October 2004 to September 2017) only the coefficients were fed into the models and the models attempted to predict the ERAi temperature anomaly at the chosen grid point. In this case, the

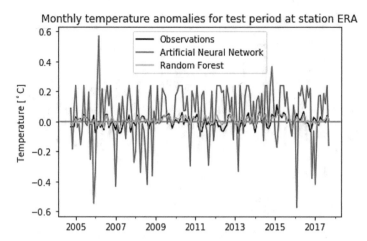

Fig. 4 Monthly temperature anomalies at a grid point in the ERAi (coarse-scale) climatology. NOTE: "Observations" are actually ERAi reanalysis anomalies at the grid point

training and testing data sets do not intersect, but the "observation" is a reanalysis value, making this second experiment another idealized scenario.

Figure 4 is similar to Fig. 2 but shows the performance of the models in predicting reanalysis temperature anomaly at a grid point of the coarse-scale climate data set. We see that the RF results capture the amplitude of anomalies much better than the ANN model implementation. Ultimately, both idealized tests (experiment 1 and 2) suggest that the RF has a higher upper bound of performance attainable compared to the ANN.

3.3 Experiment 3: Predicting Weather Station Data

In the third experiment, the RF and ANN algorithms were trained on the months October to December, 1979 through 2004, to make predictions at weather stations for the October to December season during the years 2005 through 2017. In this experiment, all training and testing data come from the autumn months, and the training and testing data sets are disjoint.

Each panel of Fig. 5 is similar to Fig. 3, showing ML model-downscaled anomaly (labeled 'predictions') as a function of observations for six different stations on flat land (see Fig. 6), where the machine learning models were most successful, for the October through December season. Despite the differences between RF and ANN when applied to the two idealized scenarios, in this experiment they appear to perform similarly.

The final test also compared the results of the RF and ANN models with the simplest form of machine learning: linear regression (LR). Figure 7 shows a comparison of the three models using ANN (blue), RF (orange), and LR (purple) for a location

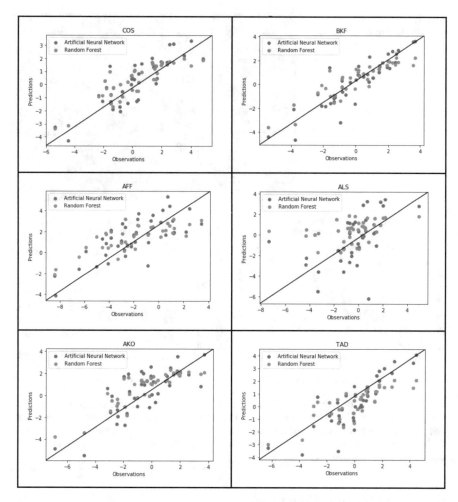

Fig. 5 Predictions of temperature anomalies (°C) at Colorado weather stations for October through December

that is flat (left panel) and one in complex terrain (right panel). In this example, all three techniques perform better for flat ground than for the mountainous region, indicating that the machine learning techniques struggle in complex topography. Downscaling the temperature on flat terrain is likely easier because it lacks the complicated temperature heterogeneities of complex topography.

Fig. 6 Location of seven
weather stations in Colorado

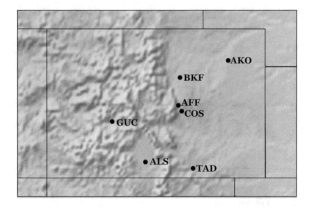

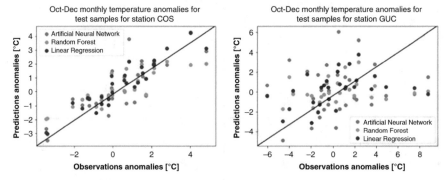

Fig. 7 Comparison of three ML techniques for a location in flat terrain and one location in complex terrain. Predicted monthly temperature anomalies vs. observed monthly temperature anomalies for October, November and December at a location on flat terrain (COS/Colorado Springs, left) and mountainous terrain (GUC/Gunnison, right). See Fig. 6 for additional information about these geographic locations

4 Discussion and Conclusions

During a four-day workshop, the research team tested the use of machine learning algorithms in downscaling ERA-Interim reanalysis data to estimate monthly temperature averages at weather stations. We found the machine learning methods that we implemented (RF and ANN) performed similarly to linear regression, and that all three methods had more skill at downscaling temperatures for weather stations on smoother terrain.

We can contrast these preliminary findings with those of previous studies (see Sect. 2): Coulibaly et al. [23] and Duhan and Pandey [24] found ANN performed similarly to linear methods, while Goyal and Ojha [25] found ANN superior to multilinear regression (MLR) when downscaling monthly mean maximum and minimum temperatures. Weichert and Bürger [22] found modest improvements through using ANN compared to more traditional techniques for downscaling daily average temperatures while Samadi et al. [21] found ANN less suitable than a linear method for the same task. Perhaps relevant to our third experiment, in a daily precipitation downscaling study Tolika et al. [27] found ANN was successful in the winter and spring seasons, but not autumn. Holden et al. [26] found random forests to be a viable method for downscaling minimum temperature; Pang et al. [20] found RF to perform better than ANN or MLR for downscaling average daily temperatures.

The fact that we saw added value in RF compared to ANN in the idealized scenarios is in line with the findings of previous studies, in which ANN has had less uniformly positive outcomes than has RF. However, our finding that ANN and RF performed similarly to each other, and to linear regression, in the third experiment is more surprising. Perhaps the most likely explanation for the unsatisfactory performance of the ML algorithms is that they were used by inexperienced researchers under time constraints and with little tuning. Configuring RF or ANN involves specifying parameters like number of nodes and layers (for ANN) or numbers and depths of trees (for RF); the user must also decide what sort of features to provide and how many. In the short duration of the workshop we were unable to thoroughly explore this parameter space. Also, the lack of quality control on the observations (as discussed in Sect. 2.1) may have been a compounding factor. Another possibility is that this particular problem is dominated by linear effects (as hinted at in our second experiment, in which the ERAi value to be predicted is exactly linear in a long enough series of features) when the only inputs are temperature fields. Another possibility is limitations imposed by the number of principal components retained and ability of coarse reanalysis temperatures in representing local-scale temperatures; whereas other studies used a suite of model parameters and harnessed ML's ability to select predictors, we attempted downscaling using just a small number of features.

Many other facets of using machine learning to downscale climate data were of interest but were not investigated due to the limited time available during the workshop. As mentioned, we had little time to calibrate the ML models. We also used only one variable—temperature—from the reanalysis data set and we never tested the methods using climate projections such as CESM-LE. While we experimented with temperature data as output, we were also interested in applying the methods to other variables like precipitation or snow water equivalent. Both present even more challenges for downscaling but also represent critical applications to resource managers and communities.

Acknowledgements Partial funding for this research was provided by NOAA Award NA19OAR4320073. The views, opinions, and findings contained in this report are those of the authors and should not be construed as an official National Oceanic and Atmospheric Administration or U.S. government position, policy, or decision.

References

1. Lucas-Picher, P., Laprise, R. and Winger, K., 2017. Evidence of added value in North American regional climate model hindcast simulations using ever-increasing horizontal resolutions. Climate Dynamics, 48(7-8), pp.2611-2633.
2. Maraun, D., Wetterhall, F., Ireson, A.M., Chandler, R.E., Kendon, E.J., Widmann, M., Brienen, S., Rust, H.W., Sauter, T., Themeßl, M. and Venema, V.K.C., 2010. Precipitation downscaling under climate change: Recent developments to bridge the gap between dynamical models and the end user. Reviews of Geophysics, 48(3).
3. Meehl, G.A., Zwiers, F., Evans, J., Knutson, T., Mearns, L. and Whetton, P., 2000. Trends in extreme weather and climate events: issues related to modeling extremes in projections of future climate change. Bulletin of the American Meteorological Society, 81(3), pp.427-436. https://doi.org/10.1175/1520-0477(2000)081<0427:TIEWAC>2.3.CO;2
4. Wuebbles, D., Meehl, G., Hayhoe, K., Karl, T.R., Kunkel, K., Santer, B., Wehner, M., Colle, B., Fischer, E.M., Fu, R. and Goodman, A., 2014. CMIP5 climate model analyses: climate extremes in the United States. Bulletin of the American Meteorological Society, 95(4), pp.571-583. https://doi.org/10.1175/BAMS-D-12-00172.1
5. Maurer, E.P., Brekke, L., Pruitt, T. and Duffy, P.B., 2007. Fine-resolution climate projections enhance regional climate change impact studies. Eos, Transactions American Geophysical Union, 88(47), pp.504-504.
6. Fischbach, Jordan R., Kyle Siler-Evans, Devin Tierney, Michael Wilson, Lauren M. Cook, and Linnea Warren May, Robust Stormwater Management in the Pittsburgh Region: A Pilot Study. Santa Monica, CA: RAND Corporation, 2017. RR-1673-MCF
7. Abatzoglou, J.T. and Brown, T.J., 2012. A comparison of statistical downscaling methods suited for wildfire applications. International Journal of Climatology, 32(5), pp.772-780.
8. Wood, A.W., Leung, L.R., Sridhar, V. and Lettenmaier, D.P., 2004. Hydrologic implications of dynamical and statistical approaches to downscaling climate model outputs. Climatic change, 62(1-3), pp.189-216.
9. Chegwidden, O. S., Nijssen, B., Rupp, D. E., Arnold, J. R., Clark, M. P., Hamman, J. J., et al. (2019). How do modeling decisions affect the spread among hydrologic climate change projections? Exploring a large ensemble of simulations across a diversity of hydroclimates. Earth's Future, 7, 623–637. https://doi.org/10.1029/2018EF001047
10. Vano, J.A., Arnold, J.R., Nijssen, B., Clark, M.P., Wood, A.W., Gutmann, E.D., Addor, N., Hamman, J. and Lehner, F., 2018. DOs and DON'Ts for using climate change information for water resource planning and management: guidelines for study design. Climate Services. DOI: https://doi.org/10.1016/j.cliser.2018.07.002
11. Hao, K.: What is machine learning? MIT Technology Review. https://www.technologyreview.com/s/612437/what-is-machine-learning-we-drew-you-another-flowchart/(2018). Accessed 6 Sep 2019
12. Harlarka, R.: Choosing the right machine learning algorithm. https://hackernoon.com/choosing-the-right-machine-learning-algorithm-68126944ce1f(2018). Accessed 6 Sep 2019
13. Roßbach, P.: Neural Networks vs. Random Forests – Does it always have to be Deep Learning? https://blog.frankfurt-school.de/neural-networks-vs-random-forests-does-it-always-have-to-be-deep-learning/(2018). Accessed 6 Sep 2019
14. Brenowitz, N.D. and Bretherton, C.S., 2018. Prognostic validation of a neural network unified physics parameterization. Geophysical Research Letters, 45(12), pp.6289-6298. https://doi.org/10.1029/2018GL078510
15. Crawford, J., Venkataraman, K. and Booth, J., 2019. Developing climate model ensembles: A comparative case study. Journal of hydrology, 568, pp.160-173.
16. Dee, D.P., Uppala, S.M., Simmons, A.J., Berrisford, P., Poli, P., Kobayashi, S., Andrae, U., Balmaseda, M.A., Balsamo, G., Bauer, D.P., and Bechtold, P., 2011. The ERA-Interim reanalysis: Configuration and performance of the data assimilation system. Quarterly Journal of the royal meteorological society, 137(656), pp.553-597. https://doi.org/10.1002/qj.828

17. Kay, J. E., Deser, C., Phillips, A., Mai, A., Hannay, C., Strand, G., Arblaster, J., Bates, S., Danabasoglu, G., Edwards, J., Holland, M. Kushner, P., Lamarque, J.-F., Lawrence, D., Lindsay, K., Middleton, A., Munoz, E., Neale, R., Oleson, K., Polvani, L., and M. Vertenstein (2015), The Community Earth System Model (CESM) Large Ensemble Project: A Community Resource for Studying Climate Change in the Presence of Internal Climate Variability, Bulletin of the American Meteorological Society, 96, pp.1333-1349. doi: https://doi.org/10.1175/BAMS-D-13-00255.1

18. Tyndall, D. P. and Horel, J.D., 2013. Impacts of mesonet observations on meteorological surface analyses. Weather and Forecasting, 28(1), pp.254-269.

19. Géron, A., 2017. Hands-on machine learning with Scikit-Learn and TensorFlow: concepts, tools, and techniques to build intelligent systems. O'Reilly Media, Inc.

20. Pang, B., Yue, J., Zhao, G. and Xu, Z., 2017. Statistical downscaling of temperature with the random forest model. Advances in Meteorology, 2017.

21. Samadi, S., Wilson, C.A. and Moradkhani, H., 2013. Uncertainty analysis of statistical downscaling models using Hadley Centre Coupled Model. Theoretical and applied climatology, 114(3-4), pp.673-690.

22. Weichert, A. and Bürger, G., 1998. Linear versus nonlinear techniques in downscaling. Climate Research, 10(2), pp.83-93.

23. Coulibaly, P., Dibike, Y.B. and Anctil, F., 2005. Downscaling precipitation and temperature with temporal neural networks. Journal of Hydrometeorology, 6(4), pp.483-496.

24. Duhan, D. and Pandey, A., 2015. Statistical downscaling of temperature using three techniques in the Tons River basin in Central India. Theoretical and applied climatology, 121(3-4), pp.605-622.

25. Goyal, M.K. and Ojha, C.S.P., 2012. Downscaling of surface temperature for lake catchment in an arid region in India using linear multiple regression and neural networks. International Journal of Climatology, 32(4), pp.552-566.

26. Holden, Z.A., Abatzoglou, J.T., Luce, C.H. and Baggett, L.S., 2011. Empirical downscaling of daily minimum air temperature at very fine resolutions in complex terrain. Agricultural and Forest Meteorology, 151(8), pp.1066-1073.

27. Tolika, K., Maheras, P., Vafiadis, M., Flocas, H.A. and Arseni-Papadimitriou, A., 2007. Simulation of seasonal precipitation and raindays over Greece: a statistical downscaling technique based on artificial neural networks (ANNs). International Journal of Climatology: A Journal of the Royal Meteorological Society, 27(7), pp.861-881.

28. Kluyver, T., Ragan-Kelley, B., Pérez, F., Granger, B.E., Bussonnier, M., Frederic, J., Kelley, K., Hamrick, J.B., Grout, J., Corlay, S. and Ivanov, P., 2016, May. Jupyter Notebooks-a publishing format for reproducible computational workflows. In ELPUB (pp. 87-90).

Approaches to Analyzing the Vulnerability of Community Water Systems to Groundwater Contamination in Los Angeles County

Michelle E. Miro, Kelsea Best, Nur Kaynar, Rachel Kirpes, and Aisha Najera Chesler

Abstract Groundwater resources are increasingly drawn on as means to buffer surface water shortages during droughts as well as to improve the Los Angeles region's reliance on local, rather than imported, water sources. However, the Los Angeles region is home to a legacy of contamination that threatens the quality and safety of groundwater as a drinking water resource. As utilities and other water management entities in the Los Angeles region look to increase their reliance on groundwater resources, a comprehensive understanding of which community water systems may be vulnerable to contamination can help planners, policy makers and regulators support these communities. This paper details the objectives, process and lessons learned from a workshop-based research project that examined the spatial extent of groundwater contamination in Los Angeles County's groundwater basins. Our team of researchers cleaned and processed multiple geospatial datasets and utilized logistic regression and machine learning methods to predict which community drinking water systems are particularly vulnerable to groundwater contamination.

M. E. Miro (✉) · A. Najera Chesler
RAND Corporation, Santa Monica, CA, USA
e-mail: mmiro@rand.org; anajerac@rand.org

K. Best
Vanderbilt University, Nashville, TN, USA
e-mail: kelsea.b.best@vanderbilt.edu

N. Kaynar
University of California, Los Angeles, Los Angeles, CA, USA
e-mail: snurkaynar@gmail.com

R. Kirpes
University of Michigan, Ann Arbor, MI, USA
e-mail: rmkirpes@umich.edu

© The Author(s) and the Association for Women in Mathematics 2020
M. Lee, A. Najera Chesler (eds.), *Research in Mathematics and Public Policy*, Association for Women in Mathematics Series 23, https://doi.org/10.1007/978-3-030-58748-2_2

1   Introduction

Groundwater resources, water found below the Earth's surface in soils and rock formations, constitute either whole or part of the water supply for about 85% of water users in California [1]. Dependence on groundwater resources is likely to grow, particularly in some parts of the state, as more frequent drought and increases in variability of precipitation decrease the availability of surface water supplies [2]. However, much of the state's groundwater resources face significant contamination—from both natural geologic contaminants and from a legacy of industrial and agricultural production [3–5]. Within the state, 680 community water systems (CWS) that collectively serve about 21 million people rely either fully or in part on contaminated groundwater [5]. Of these, 265 have distributed drinking water that violated drinking water quality standards [5].

In Los Angeles County (LA County) over 228 CWS provide drinking water to between 25 to four million LA area residents [6]. 89 of these rely on contaminated groundwater [5, 6]. Many of the systems that are fully dependent on groundwater resources are small CWS located in disadvantaged communities; these are often cited as the most vulnerable to contamination [6, 7]. The characteristics of such a vast number of water systems in LA County also vary by their governance arrangements, technical, managerial and financial capacities, community socio-economics, infrastructure age and condition, and other factors. Each of these factors impacts the quality and safety of drinking water for the system's residents.

As utilities and other water management entities in the Los Angeles region look to increases their reliance on groundwater resources to become more resilient in the face of more frequent drought and a changing climate [2], interdisciplinary research is needed to characterize why and how CWS are violating or may be likely to violate Safe Drinking Water Act (SDWA) standards.

The SDWA protects the quality of drinking water distributed by CWS to communities. Under the authority of the SDWA, the United States Environmental Protection Agency (US EPA) sets health-related drinking water quality standards, known as maximum contamination levels (MCLs) for drinking water [8]. An MCL is the highest level of a contaminant that is allowed in drinking water; MCLs are set by US EPA for 88 regulated contaminants found in public drinking water [8]. In California, the California Environmental Protection Agency (CalEPA) implements and administers the SDWA, including tracking compliance with SDWA standards and issuing enforcement actions when CWS are in violation of these standards.

The goal of this research is to identify which community water systems in LA County are most vulnerable to providing unsafe drinking water due to groundwater contamination, and to offer insight into which factors contribute to this vulnerability. This information can help policymakers and planners target specific systems, as well as better tailor the type of support needed, with a goal of improving the health and safety of LA County residents.

The intent of the work presented in this paper is to provide an overview of the objectives, process and lessons learned of this research. While most journal articles articulate final research findings, this proceedings paper instead shows more details on research team formulation, timing, data processing, and other methodological considerations. The research team hopes such a description is useful for other data-heavy, quick-turn research projects carried out in groups.

This article details this effort, including a discussion of workshop objectives (Sect. 2), a description of data cleaning and processing (Sect. 3), an overview of details on methods formulation and selection (Sect. 4) and a final section that discusses the lessons learned and future work (Sect. 5). A more comprehensive literature review, methods discussion and the final results of the research are detailed in a companion piece that is in preparation at the time of writing.

2 Workshop Objectives

In January 2019, 41 women with backgrounds in both public policy and science, technology, engineering and mathematics (STEM) joined a 4-day workshop, Women in Mathematics and Public Policy (WPOL), at the University of California, Los Angeles (UCLA) Institute for Pure and Applied Mathematics (IPAM). The workshop was designed to achieve two major goals: (1) to provide a space for women to expand their professional networks; and (2) to address a key policy problem through the application of STEM methodologies. In particular, the workshop focused on six policy problems—three related to climate change and environmental management and three related to cybersecurity.

Within the workshop, the team organized to address the policy problem detailed in this paper included: a water resources engineer, an industrial engineer, an environmental scientist, an atmospheric chemist, and a mathematician. Through the course of the workshop, the team worked together to collectively address the policy problem of groundwater contamination of LA County. Each team member strove to broaden her disciplinary perspectives, learn new methodologies, and build her research network.

3 Data Cleaning and Processing

To build a full database of data relevant to the vulnerability of community water systems to groundwater contamination, the research team compiled groundwater quality monitoring data, the number of contaminant sources in the service zones of community water systems, the size of community water systems, their percent reliance on groundwater and past violations of SDWA standards, see Table 1. The majority of these datasets required processing to align them by location and compute water quality exceedances above regulatory limits. In addition, raw monitoring data had missing data, typos or computer errors in values (e.g. physically infeasible values) and required cleaning.

Table 1 Datasets and sources

	Dataset	Source
Water quality	Contaminant levels in monitoring wells	[9]
Contaminant sources	Location, status of environmental hazard sites	[10]
	Location, condition of underground storage tanks	[10]
	Location of waste disposal sites (e.g. landfills)	[10]
	Percent reliance on groundwater	[11]
CWS characteristics	History of violations of SDWA	[12]
	System size (by population served)	[13]
	Service area	[11]

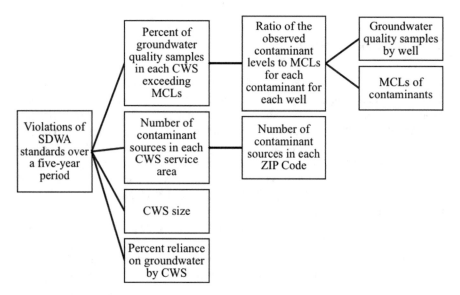

Fig. 1 Flow chart of dataset processing

Figure 1, below, shows an overview of the goals of the data cleaning and processing tasks, which included quantifying the percent of groundwater water quality samples that violated SDWA standards, calculating the number of contaminant sources in CWS service zones, and formatting data on system size and percent reliance on groundwater.

The State of California's Groundwater Ambient Monitoring and Assessment (GAMA) Program includes groundwater quality data on dozens of contaminants across the state [9]. Sixty different documented contaminant types were found in Los Angeles County groundwater; of these, arsenic was the most frequent. With this dataset, the research team calculated by how much water quality measurements were above or below MCLs. The research team then calculated the percentage of samples in which the contamination level of a given chemical exceeded the corresponding

MCL. Each exceedance was georeferenced and included as a variable in the study database.

The size of a given community water system was estimated by using the population served by each system as a proxy. This data, as well as that containing the geographic boundaries of each CWS, was obtained from California's State Water Resources Control Board (SWRCB) Department of Drinking Water [13] in a shapefile format for use with Geographic Information System (GIS) software. This data was extracted by converting the data from vector to raster format, georeferencing the raster and saving the data into the study database. The boundaries of each CWS were also saved in GIS for later mapping.

Data on the percent of water supply obtained from groundwater for each CWS was also included in the database. This data was collected as a part of a UCLA study [11].

To characterize the relationship between groundwater quality and likely sources of contamination, data on environmentally hazardous sites compiled by the State of California's GeoTracker database was included [10]. This dataset is categorized by type of clean-up site (National Priority List sites, brownfields, etc.) and the status of clean-up. Additional data was downloaded on the locations of underground storage tanks and their condition (e.g. leaky, not leaky), as well as the locations of waste disposal sites (landfills). All locations were georeferenced by ZIP Code, and a total count of sites within the boundaries of each ZIP Code was calculated. The ZIP Code counts of sites with likely contamination sources were then merged with the boundaries of CWS to obtain the number of contamination sources in each CWS area.

Finally, historical data on MCL violations of the SDWA was downloaded from California's SWRCB [12] for 2013–2016.[1]

4 Method Selection and Description

This research is targeted at identifying which community water systems in LA County are most vulnerable to providing unsafe drinking water due to groundwater contamination, and to offer insight into which factors contribute to this vulnerability. To do so, we make the assumption that the vulnerability of community water systems to violating SDWA standards is based upon the magnitude of ground-water contamination, the number of contaminant sources in the service zones of community water systems, the size of community water systems, their percent reliance on groundwater and past violations of SDWA standards (see Fig. 2). We do not specifically account for the level of treatment in each CWS and make the assumption that these factors we do account for enhance the vulnerability of CWS

---

[1]Compliance data from 2017 was not available at the time of analysis, and data from 2012 and earlier was not available in tabular format.

Fig. 2 Modeled relationship between predictor variables (right) and output (left)

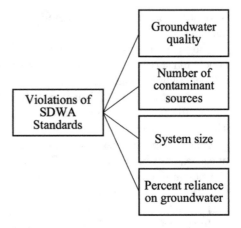

to SDWA violations, regardless of the level of treatment. Further, obtaining detailed information on the specific treatment processes and technical capacity of individual CWS was infeasible for the scope of this study.

To develop a numerical model that links the input variables to the output, the interdisciplinary team chose to use a data-driven approach employing statistical learning to predict violations of SDWA standards in LA County. Statistical learning, also referred to commonly as machine learning, describes a range of methods to identify patterns in datasets and use these patterns to construct statistical models. This quality of statistical learning may be especially useful in the context of complex, interdisciplinary problems, as these methods are largely agnostic to theory. Previous work has shown that statistical learning methods can be used to predict groundwater contamination [14–19].

Though there are many types of statistical learning algorithms, ranging in complexity and interpretability, traditional approaches to predicting groundwater contamination have depended heavily on multivariate logistic regression or linear regression with varying success [20–23]. However, when prediction is the goal of an analysis, other machine learning algorithms may be superior to linear models. Two such methods include artificial neural networks (ANN) and random forests (RF).

In this work, ANN and RF were selected for analysis because of their high predictive ability. Another aim of this study was to compare these two methods to each other and to the more traditional logistic regression method. In each of these three approaches, the outcome variable modeled for prediction was a binary variable representing whether or not a CWS may violate the SDWA, with the predictor variables highlighted in Fig. 1. Therefore, each model received the same data with the goal of predicting the same outcome, and what varied was the method of statistical learning (ANN, RF, or logistic regression).

Artificial neural networks (ANN) are flexible, statistical learning networks designed to mimic the neurons in a human brain. They are comprised of nodes, or neurons, that are linked together through weighted connections that in their simplest form multiply the value of the node by the weight of the connection. ANNs may have many nodes (and thus various weights on connections between those nodes) and multiple layers of nodes that ultimately connect multiple inputs to the output. Layers pass information from input to output through activation functions that can turn on when the preceding node value is above a given threshold or transform the value before passing information along. The weights which connect nodes are calibrated, or "trained" with machine learning methods. Once trained, ANNs can predict output at different points in time or space, recognize patterns or categorize output. They represent a rather fast and efficient method for characterizing non-linear relationships and deriving predictions.

Random forest (RF) models are an ensemble method of decision trees, representing a subset of statistical learning known as tree-based methods. Tree-based methods, including random forests, can be used for classification of discrete outcome variables, as in this research, or regression of continuous variables. They are especially powerful when there are nonlinearities in the data, or the relationships between variables are complex, as is likely the case with the groundwater contamination. RF models work by fitting many decision trees (where the number of trees may be specified by the modeler), in which each decision tree uses a random subset of predictor variables at each split of the tree. The ultimate model prediction is then obtained by averaging across all of the predictions from the individual decision trees, which is why this method is called a "forest". By averaging across many individual trees, RF models are able to obtain very high predictive accuracy while avoiding challenges of overfitting data. RF models are also very flexible and can take different combinations of data inputs without requiring scaling or manipulation of data.

Logistic regression is a widely used predictive analysis method in which the dependent variable is categorical. Logistic regression has a wide application in multiple scientific disciplines, including biology, healthcare, and environmental sciences.

Another especially useful element of ANN, RF and logistic regression models is their ability to assess the relative importance of each input variable. For ANN, Garson's method can be utilized to quantify the relative contribution of each input parameter to the output [24]. For RF, variable importance, a measure of a variable's contribution to the model performance, can be calculated. In logistic regression, the coefficients derived from the model can be used to calculate the odds ratio (by taking the antilog of the coefficient), which in this case can be interpreted as the increased odds of SDWA violations given one unit increase in the given independent variable. Though these measures of importance cannot be meaningfully compared across the three different methods, they can be used to rank variable contribution to the overall model within an analysis, as shown in this work. In this way, variable importance measures can shed light on which variables are most predictive of the outcome, while accounting for nonlinearity and complexity in the data.

With the goal of predicting which CWS may violate the SDWA, ANNs, RF and logistic regression models were trained using the database developed for this study. Specifically, known violations of SDWA were used as the dependent variable and system size, level of groundwater contamination, percent reliance on groundwater and the number of contamination sources within a CWS service area were modeled as independent variables. Training data was obtained by randomly selecting 70% of the total data and the remaining 30% of the data was used to test the model. The ANN and RF models performed well, showing low errors in training, validation and testing. The logistic regression found no significant relationship between the dependent variable, whether or not a Community Water System has violated the Safe Drinking Water Act, and the independent variables. Further details and the full results of this analysis will be published in a companion article.

5   Discussion

The research team derived a set of key lessons learned from this exercise. Principally, one important value add of the workshop was a dedicated amount of time to focus on a single research project, discuss research goals, troubleshoot questions and carry out analyses. Daily work time with ample space and resources allowed the team to produce research results efficiently. Interaction with other research teams pursuing similar environmental science topics also allowed the team to refine the choice of methodologies and more clearly structure results. Another benefit of the workshop format for this project was the team it produced. The research team brought expertise from five distinct disciplines, which yielded creative application of methods, new ways of solving problems and different perspectives on both the technical and policy aspects of the project. Finally, this proceedings piece illustrates the amount of time and effort taken to clean, process and format the input data. Because ANN, RF and logistic regression are data-driven methodologies, the quality, quantity and choice of data is of utmost importance to the value of output the models produce. In a real-world policy application of such methods, necessary data must be clearly defined and then carefully reviewed and utilized prior to applying statistical learning methods.

Future research and analysis steps for this work could incorporate socioeconomic data, such as a social vulnerability index (SVI) into the analysis as a predictor. This would enable a deeper examination of SVI as a factor that may make a given CWS more vulnerable to contamination and thus more likely to violate SDWA standards. A more comprehensive and technically-sound treatment of groundwater contamination and contaminant transport could support an analysis that is targeted at identifying which CWS are vulnerable to contamination and when contaminant plumes may reach their groundwater supply source wells. Such an analysis would likely involve more complete physics-based groundwater quality modeling (such as in [4]). Finally, the impact of this work could be extended by presenting its findings to drinking water regulatory authorities in the State of California.

References

1. Chappelle, C., Hanak, E., & Harter, T.: Groundwater in California. https://www.ppic.org/wp-content/uploads/JTF_GroundwaterJTF.pdf (2018). Accessed 30 January 2019
2. Diffenbaugh, N. S., Swain, D. L., Touma, D.: Anthropogenic warming has increased drought risk in California. Proceedings of the National Academy of Sciences, 112(13), 3931-3936 (2015). https://doi.org/10.1073/pnas.1422385112
3. Helperin, A., Beckman, D. & Inwood, D.: California's Contaminated Groundwater: Is the State Minding the Store? Natural Resources Defense Council. https://www.nrdc.org/sites/default/files/ccg.pdf (2001). Accessed 30 January 2019
4. Ponti, D. J., Wagner, B. J., Land, M., Landon, M. K.: Characterization of potential transport pathways and implications for groundwater management near an anticline in the Central Basin area, Los Angeles County, California. U.S. Geological Survey Open-File Report, 1087 (2014). https://doi.org/10.3133/ofr20141087
5. State Water Resources Control Board: Communities that Rely on a Contaminated Groundwater Source for Drinking Water. Report to the Legislature. https://www.waterboards.ca.gov/gama/ab2222/docs/ab2222.pdf (2013). Accessed 30 January 2019
6. DeShazo, J.R., Pierce, G., McCann, H.: Los Angeles Community Water Systems: Atlas and Policy Guide. UCLA Luskin Center for Innovation. https://innovation.luskin.ucla.edu/sites/default/files/Water_Atlas_0.pdf (2015). Accessed 30 January 2019
7. Sacramento State: Assessing Groundwater Contamination Risks in California's Disadvantaged Communities. Sacramento State Office of Water Programs. http://www.owp.csus.edu/grid/ (2018). Accessed 30 January 2019
8. United States Environmental Protection Agency: Summary of the Safe Drinking Water Act. Laws and Regulations. https://www.epa.gov/laws-regulations/summary-safe-drinking-water-act (2019). Accessed 30 January 2019
9. State Water Resources Control Board: Groundwater Ambient Monitoring and Assessment (GAMA) Program. Division of Water Quality. https://www.waterboards.ca.gov/water_issues/programs/gama/online_tools.html (2019b). Accessed 30 January 2019
10. State Water Resources Control Board: GeoTracker. http://geotracker.waterboards.ca.gov/datadownload (2019a). Accessed 30 January 2019
11. Los Angeles Water Hub: University of California Center for Sustainable Communities. http://waterhub.ucla.edu/ (2019). Accessed 30 January 2019
12. California Department of Drinking Water: Annual Compliance Report. State Water Resources Control Board. https://www.waterboards.ca.gov/drinking_water/certlic/drinkingwater/Publications.html#compliance (2019). Accessed 30 January 2019
13. California Department of Drinking Water: California Community Water Systems Inventory Dataset. State Water Resources Control Board. https://catalog.data.gov/dataset/california-community-water-systems-inventory-dataset-2010 (2010). Accessed 30 January 2019.
14. Choubin, B., Solaimani, K., Habibnejad Roshan, M., Malekian, A.: Watershed classification by remote sensing indices: A fuzzy c-means clustering approach. Journal of Mountain Science, 14(10), 2053–2063 (2017). https://doi.org/10.1007/s11629-017-4357-4
15. Choubin, B., Darabi, H., Rahmati, O., Sajedi-Hosseini, F., Kløve, B.: River suspended sediment modelling using the CART model: A comparative study of machine learning techniques. Science of The Total Environment, 615, 272–281 (2018). https://doi.org/10.1016/j.scitotenv.2017.09.293
16. Khalil, A., Almasri, M. N., McKee, M., Kaluarachchi, J. J.: Applicability of statistical learning algorithms in groundwater quality modeling. Water Resources Research, 41(5) (2005). https://doi.org/10.1029/2004WR003608
17. Rodríguez-Lado, L., Sun, G., Berg, M., Zhang, Q., Xue, H., Zheng, Q., Johnson, C. A.: Groundwater Arsenic Contamination Throughout China. Science, 341(6148), 866–868 (2013). https://doi.org/10.1126/science.1237484

18. Sajedi-Hosseini, F., Malekian, A., Choubin, B., Rahmati, O., Cipullo, S., Coulon, F., Pradhan, B.: A novel machine learning-based approach for the risk assessment of nitrate groundwater contamination. Science of The Total Environment, 644, 954–962 (2018). https://doi.org/10.1016/j.scitotenv.2018.07.054

19. Singh, K. P., Gupta, S., Mohan, D.: Evaluating influences of seasonal variations and anthropogenic activities on alluvial groundwater hydrochemistry using ensemble learning approaches. Journal of Hydrology, 511, 254–266 (2014). https://doi.org/10.1016/j.jhydrol.2014.01.004

20. Bretzler, A., Lalanne, F., Nikiema, J., Podgorski, J., Pfenninger, N., Berg, M., Schirmer, M.: Groundwater arsenic contamination in Burkina Faso, West Africa: Predicting and verifying regions at risk. Science of The Total Environment, 584–585, 958–970 (2017). https://doi.org/10.1016/j.scitotenv.2017.01.147

21. Cao, H., Xie, X., Wang, Y., Pi, K., Li, J., Zhan, H., Liu, P.: Predicting the risk of groundwater arsenic contamination in drinking water wells. Journal of Hydrology, 560, 318–325 (2018). https://doi.org/10.1016/j.jhydrol.2018.03.007

22. Li, Y., Wang, D., Liu, Y., Zheng, Q., Sun, G.: A predictive risk model of groundwater arsenic contamination in China applied to the Huai River Basin, with a focus on the region's cluster of elevated cancer mortalities. Applied Geochemistry, 77, 178–183 (2017). https://doi.org/10.1016/j.apgeochem.2016.05.003

23. Mfumu Kihumba, A., Ndembo Longo, J., Vanclooster, M.: Modelling nitrate pollution pressure using a multivariate statistical approach: the case of Kinshasa groundwater body, Democratic Republic of Congo. Hydrogeology Journal, 24(2), 425–437 (2016). https://doi.org/10.1007/s10040-015-1337-z

24. Garson, G. D.: Interpreting neural-network connection weights. AI expert, 6(4), 46-51. (1991)

The Measurement of Disaster Recovery Efficiency Using Data Envelopment Analysis: An Application to Electric Power Restoration

Priscillia Hunt and Kelly Klima

Abstract Many firms are involved in rebuilding a region's critical infrastructure after a disaster. One question that follows is how efficient were firms in delivering restoration services? The evaluation approach proposed in this paper builds off Reilly et al. (Reliability Engineering & System Safety 152: 197–204, 2016) by considering disaster severity to assess efficiency. We expand on the approach by adjusting limitations on the upper bound of efficiency. This paper similarly provides an application to recent post-hurricane restoration activities across different electric power companies in the states and territories of the United States.

1 Introduction

The North American Electric Reliability Council (NERC) requires[1] utilities to report large outages, and the NERC data shows adverse weather is a leading cause of power outages. Wind and rain events are the most common cause of outages, resulting in approximately one third of the outages reported to NERC [1]. Hurricanes and tropical storms cause approximately 10% of the outages reported to NERC, and amongst weather causes, hurricanes and tropical storms have the highest average number of customers without power per incident (approximately 900,000

The original version of this chapter was revised. A correction to this chapter is available at https://doi.org/10.1007/978-3-030-58748-2_7

---

[1] Federal Power Act, Section 215(e); 18 C.F.R. §39.7.

P. Hunt (✉)
RAND Corporation, Santa Monica, CA, USA

IZA, Bonn, Germany
e-mail: phunt@rand.org

K. Klima
RAND Corporation, Santa Monica, CA, USA
e-mail: kklima@rand.org

© The Author(s) and the Association for Women in Mathematics 2020, corrected   29
publication 2021
M. Lee, A. Najera Chesler (eds.), *Research in Mathematics and Public Policy*, Association for
Women in Mathematics Series 23, https://doi.org/10.1007/978-3-030-58748-2_3

customers). Indeed, seven of the fifteen largest blackouts prior to 2009 were caused by hurricanes [1]. Given that weather-related outages cause at least $20 billion to $55 billion in direct losses annually in USD 2012 [2], there is a need for research to understand the level of efficiency in restoring electrical power after an outage and the factors affecting efficiency values.

This study aims to assess the performance of utility companies restoring power after moderate to extreme hurricanes in U.S. states and territories, and to explore the factors that can influence the efficiency of power restoration. We are not the first to think about this. Reilly et al. [3] developed a process for evaluating restoration of electric power after hurricanes. The authors used a well-known approach, data envelopment analysis (DEA), to compare efficiency of companies contracted to restore power after hurricanes in the continental U.S. between 1996 and 2012. As the authors explained, however, the performance metrics applied in their study were for illustrative purposes, and other metrics may be more appropriate. Furthermore, the authors used the three-stage DEA by Gorman and Ruggiero [4], which has the benefit of accounting for heterogeneity in the environment in which decision makers operate, but the limitation that the most efficient restorations in the dataset cannot be compared to any other hurricane; in other words, they were all ranked the same.

Therefore, this paper makes three contributions to the literature. First, we considered a different DEA algorithm—"super-efficiency" three-stage DEA—that eliminates the upper bound on the technical efficiency score and provides additional information regarding the relative performance of the efficient units. Second, we tested the inclusion of an additional environmental factor to equalize the environmental context in which each utility company had to work during a restoration. Third, we added responses from more recent hurricanes that have been considered relatively severe, including five different utilities for Hurricane Katrina in Louisiana and one utility for Hurricane Maria in Puerto Rico.

Against this background, this study adopts the three-stage, super-efficient DEA evaluation approach [5] to estimate company performance in restoring power after moderate to extreme hurricanes. The super-efficient DEA model addresses the non-ordering problems of efficient units, and the three-stage DEA approach can eliminate the impediments of environmental factors to completing power restorations.

The following sections of this paper are as follows: Sect. 2 reviews existing studies and constructs a theoretical framework for our analysis. Section 3 presents the data and explains the methodology. Section 4 describes our analysis and examines the factors that determine post-hurricane power restoration performance. Discussions and concluding comments are provided in Sect. 5.

2 Previous Literature Assessing Power Restoration Performance

Theoretically, the lifetime of an electricity outage event follows a few steps [6]. Prior to the event, utility operators plan or prepare for the event. Then the event occurs, during which the fraction of load served is reduced. People endure a lack of electrical power while emergency workers (both in the field and in the office) assess the situation within safety limits. Then the emergency restoration begins, which can be a combination of repairing and replacing. Eventually, recovery occurs to a state where customers are energized, and since numbers of customers are always evolving, the fraction of load served may be different from the pre-storm state. In this paper, we consider the time between the event and the state where customers are energized, which we will call *time to repair*.

Given society's emphasis on reducing time to repair, there is a large body of empirical research modelling the likelihood of outages and duration of those outages (for literature reviews see Vaiman et al. [7]; Castillo [8]; Wang et al. [9]; Liu and Zhong [10]). There are three general classes of outage duration models, including regression models [11–17], regression trees [18, 19], and multivariate frameworks [20]. These studies suggested that the time to repair is affected by the model choice and the variables included which are typically a combination of infrastructure and environmental data with some models also considering crews available to effect repairs and the effect of vegetation [21].

Unfortunately, many of these models are limited in scope to one utility or service area, and do not compare restorations across state contexts. One paper applied an advanced ensemble data mining model for the entire US coastline using only publicly available data [19]. While authors predicted the occurrence of outages, they did not indicate if they were able to predict duration of outages and do not compare across regions. One exception is Reilly et al. [3]. The study uses Data Envelopment Analysis (DEA) to assess performance of utility companies following a hurricane. DEA is considered a powerful performance measurement and benchmarking tool for applications where the evaluated decision-making units (DMUs), such as utility companies, are described by activities representing real processes generating products or services and are based on a convex (or even linear) technology. DEA involves the application of linear programming techniques to estimate the performance of firms in an industry or the overall industry itself, such as agriculture, health care, transportation, education, manufacturing, power, energy and environment, communication, banking, and finance [22, 23]. A 2017 literature study found 693 articles applying the DEA approach to study performance and efficiency in the energy and environment sectors [24].

The procedure of a traditional DEA analysis is to first calculate a DMU's efficiency score based on a DMU's production of outputs to inputs and standardize the score relative to the output-to-input ratios of the highest scoring DMUs to generate a score between zero and one. The most efficient DMUs' scores are one; we later describe how we improve upon this limitation and eliminate this upper bound. Then, in order to account for environmental heterogeneity (i.e., not all firms work in the same environment), the scores are regressed on factors outside the control of firms that may influence efficiency.[2] The estimated value of the

---

[2]There are several approaches. Fried et al. [5] developed this approach, and it was further developed and shown to be efficient by Gorman and Ruggiero [4].

statistically significant coefficients and covariates tells us something about the environment conditions under which a DMU operated. Those with higher values worked in more positive environments, which improved their efficiency score, and vice versa. As a third stage, the initial efficiency score stage is re-run, but this time the standardization only occurs with other DMUs working in the same or worse environments; we show this mathematically in the Sect. 3.2. This means that if a firm receives a score of less than one, they did so because compared to other firms working in the same or worse environments, their output-to-input ratio was lower. Consequently, the DMU operating in the worst condition always scores one because it cannot be compared to any other DMU.

While many energy and environment studies apply the traditional one- and two-stage DEA method, these approaches fail to account for environmental factors outside the control of the decision-making units (DMUs) being assessed. For example, in the case of power restoration, vegetation and hurricane severity affect the speed of the restoration process, and one- and two-stage DEA approaches would not account for this in generating final efficiency values. In a three-stage DEA, the efficiency value of a DMU calculated in a first stage. In the second stage, the first-stage value is decomposed into environmental factors and all other factors. Then in the third stage, the linear programming problem is solved again to generate efficiency values but this time in groups of DMUs with a similar environment as identified in the second stage. The three-stage DEA model, which was first presented by Fried et al. [5] and further developed by Gorman and Ruggiero [4], has been shown to be an efficient model for assessing the relative efficiency for various DMUs.

The approach in this study builds off the work in Reilly et al. [3]. Authors analyzed 27 utility-storm combinations between 1996 and 2012 in the continental U.S. using a traditional three-stage DEA approach. The input considered was log of costs for all emergency and recovery work (starting with debris removal to job completion), as a proxy measure of resources used to restore power. The output measure, y, was log of customer days without power, calculated as the number of customers without power each day, summed over days until the fraction of load served on 1 day is at least 90% for multiple days. Formally, we can express y as $y = \log\left(\sum_{t=1}^{T} t * N_t\right)$, where N_t is the number of people who had power restored on day t, $t = 1$ is the first day of the disaster, and T is the number of days since the disaster occurred and the first day (of multiple days in a row) when the fraction of load served is 90%. This value of y is then inverted since the main objective is to restore power as quickly as possible to as many customers as possible and more output is better. Costs included were wire, pole, and transformer replacement costs; the cost of clean-up and restoration crews; and cost of additional support personnel. The first-stage efficiency value is based on the input and output only, as described, using an output-oriented, variable returns to scale model typically referred to as the BCC model referring to Banker et al. [25]. In the second stage, the paper tests the relationship between the first-stage efficiency value and the following environmental conditions variables: log of service area with peak wind speed over 75 miles per hour (mph), log of service area with peak wind speed over 100 mph, proportion of service area with peak wind speed over 75 mph, proportion of service area with peak wind speed over 100 mph, and log of peak number of customers without power. Only service areas with peak wind speed over 100 mph is retained in their model. In the final stage, efficiency values range from 0.13 (Hurricane Sandy) to 1.0 for five utility-storm power restoration projects. We hesitate, however, to make much of these results because the authors clearly state that the study was an initial step in understanding post-hurricane power restoration performance and should not to be used to determine whether a project was efficient or not.

3   Methodology

3.1   Framework

We frame the efficiency assessment in the context of a production function in which the objective of the power restoration company is to provide electricity to as many customers as possible as quickly as possible (Q) using labor and capital to produce electricity. The ability of workers and equipment to provide electricity to as many customers as quickly as possible is affected by the environmental context in which they are working. However, we do not assume to know the functional form of this equation, thus the production function is simply $Q = f(X)$. The utility company minimizes the customer days without power constrained by the amount of labor and capital and environmental factors beyond the control of firms.

3.2   Empirical Strategy

This study employs a DEA, which is a non-parametric technique used in the estimation of production functions and has been used extensively to estimate measures of technical efficiency in a range of industries [26]. Like the stochastic production frontiers approach, DEA estimates the maximum potential output for a given set of inputs and has primarily been used in the estimation of efficiency. Traditional DEA methods bound efficiency values between zero and one, which is a limitation of the method because all DMUs with efficiency scores of one would, therefore, be considered equally efficient [27].

To overcome the shortcomings of traditional DEA evaluation methods described earlier, this study adopts a "super-efficiency", three-stage DEA approach to evaluate the performance of utility companies that have restored power after hurricane-driven power outages. In super-efficiency DEA, each utility-storm restoration is removed from the analysis one at a time, so that if one unit out-performs what is determined to be the production frontier by the other units, that unit is assigned a value greater than one based on its radial distance to the production frontier [28]. In other words, an efficient DMU on the frontier in a normal DEA approach may be outside the technology set (i.e. efficiency greater than one) in a super-efficiency DEA.

The concept of frontier is especially important for the analysis of efficiency because efficiency is measured as the relative distance to the frontier. For example, utility companies that are technically inefficient operate at points in the interior of the frontier, while those that are technically efficient operate somewhere along the technology defined by the frontier. In a traditional DEA, a DMU is called efficient when the DEA score equals one.[3] To account for the environment affecting the ability of a utility to respond to an electrical outage, we conduct a three-stage, super-efficiency DEA. Three stages are presented as follows.

3.2.1 Stage 1: Radial Super-Efficiency DEA Method

The basic principle of a DEA is to set the evaluated units as DMUs. In our analysis, a DMU will be a utility company's response to a hurricane power outage. Each DMU shares the same inputs and outputs. Then, the efficiency value is obtained by estimating the stochastic frontier of the effective production of every DMU [29]. For each DMU, the weights of the inputs and outputs are set as variables to estimate the efficient production frontier. The extent of DMU efficiency depends on the relative distance of a DMU to the efficient production frontier.

To estimate distance to the frontier, we apply the radial super-efficiency DEA variable returns to scale, output-oriented model. The choice of input- or output-oriented models depends upon the production process characterizing the DMU. We choose an output-oriented model in which the utility companies maximize the level of output (speed with which customers are provided power) given levels of the inputs. The model is applied with only the discretionary inputs and outputs; not the environmental factors (or non-discretionary inputs). As such, we are estimating efficiency as if all the DMUs have the same environment. The 'super-efficiency' element allows us to distinguish among efficient DMUs that would all receive the same score under the traditional DEA approach. Some of the efficient DMUs (efficiency equal to one) may take on a value greater than one, but the efficiency values of DMUs that are less than efficient are not modified. We will illustrate this point concretely by providing results when using the traditional DEA model and the super-efficiency model.

Let us denote the set of DMUs as I, the set of inputs as M, and the set of outputs as N. For each DMU $i \in I$, we solve a linear program to obtain I's efficiency score. We define the following variables:

θ_i = Efficiency score of DMU $i \in I$.
X_{im} = Quantity of input $m \in M$ used by DMU $i \in I$.
Y_{in} = Quantity of output $n \in N$ produced by DMU $i \in I$.

---

[3] And all slacks are zero. If slacks do not equal zero, the DMU is said to be "weakly" efficient.

λ_{in} = Weight parameter of DMU $i \in I$.
s_{im}^- = Slack variable of input $m \in M$ for DMU $i \in I$.
s_{in}^+ = Slack variable of output $n \in N$ for DMU $i \in I$.
ε = Small (non-Archimedean) positive number.
Using these notations, we have the following model:

$$\text{Max } \theta_k + \varepsilon \left(\sum_{n \in N} s_{kn}^+ + \sum_{m \in M} s_{km}^- \right), \tag{1}$$

s.t.

$$\sum_{i \in I, i \neq k} \lambda_i Y_{in} - s_{kn}^+ = \theta_k Y_{kn} \forall n \in N, \tag{2}$$

$$\sum_{i \in I, i \neq k}^{N} \lambda_j X_{im} + s_{km}^- = X_{km} \forall m \in M, \tag{3}$$

$$\sum_{i \in I, i \neq k} \lambda_j = 1, \tag{4}$$

$$s_{km}^-, s_{kn}^+ \geq 0, \tag{5}$$

$$\lambda_j \geq 0 \forall i \in I. \tag{6}$$

Model slack, s_{kn}^+, is the amount by which inputs k can be reduced for DMU$_i$ without changing its technical efficiency, and s_{km}^-, is the augmentation in the output for DMU$_i$ that can be achieved when the inputs are also reduced consistent. The ith DMU is excluded when assessing its efficiency. Instead, as shown in Eqs. (1) and (2), the inputs and outputs of the ith DMU are replaced with the linear combination of other DMUs when assessing the efficiency of the ith DMU. In the super-efficiency model, the weight λ_j of the evaluated unit DMU$_i$ is equated to zero. Meaning, it cannot influence the efficiency score of the inefficient units, but the efficiency score of the efficient units is not limited by unity in this case. In an output-oriented model, the efficiency score, E, is the inverse (or, $1/E = \theta$).

3.2.2 Stage 2: Environmental Harshness

The prevalent method in DEA literature to find the determinants of efficiency gaps among DMUs is to use a Tobit regression analysis because the efficiency scores are censored at zero and one. Here, we use super-efficiency method, so the scores are not censored at one but they are censored at zero.

The regression uses the efficiency scores of the first stage as the dependent variables for the possible candidates of environmental variables. We linearly regress the log efficiency score estimated in stage one on the full set of nondiscretionary inputs, z. We identify statistically significant regressors, and only include those in the final regression. Therefore, a regression is estimated with the following general form:

$$\theta_i = \alpha + \sum_{j=1}^{n} \beta_j Z_{ij} + \varepsilon_i, \tag{7}$$

where β_j represents the estimator of environmental factors, z_{ij} is a matrix of environmental factors for DMUs i that can influence the translation of inputs to outputs, and θ_i is the efficiency value in the first stage of the ith DMU. Since the value of the dependent variable, θ, is censored $(0,\infty)$, we apply the Tobit regression model. This regression decomposes the index θ into the inefficiency component (ε) and environmental conditions (βz), under the assumption that inefficiency does not affect the environmental variables. Following Ruggiero [30], the environmental harshness index is calculated as $z_j^* = \sum_{j=1}^{n} \beta_i z_{ij}$, which represents the impact of environmental factors on efficiency.

3.2.3 Stage 3: Adjusted Super-Efficiency DEA Model

Thus far, the efficiency score is based on distance to the production possibility frontier under the assumption that all the DMUs face the same environment. In the context of disasters, this is clearly not the case. Some disasters are more severe and given the same level of inputs, there would be far fewer outputs in the harsher environment. To account for this, Ruggiero [31] shows, we still run the linear programming problem shown in Eqs. (1) through (6) plus the following constraint:

$$\lambda_j = 0 \forall z_i > z_{j\neq k}. \tag{8}$$

where i is the measure of harshness of the environment for DMU i obtained from the regression Eq. (7). This approach has been shown to be robust when using multiple environmental variables [3, 32].

3.3 Sample and Variables

We limit our sample to hurricanes in U.S. states and territories between 1996 and 2017 for which we have reliable data. As such, we include the utility-hurricanes in [3] and add utility responses to Hurricanes Katrina in Louisiana and Maria in Puerto Rico specifically. To conduct an empirical analysis for examining response to moderate through extreme hurricanes and to explore the potential impacts of the hurricane factors derived from the theoretical framework, we identify 33 hurricanes that occurred in U.S. states and territories between 1996 and 2017 for which we have data. A summary of the analytical data is shown in Table 1.

3.3.1 Output Variables

The primary goal of improving the length of time to restore power after an outage is to alleviate the degree of injuries, deaths, and economic losses caused by hurricanes. To reflect this objective of power restoration after hurricanes, the total number of customer days without power is adopted as the output variable of the DEA model. We then invert the measure, since the objective is to maximize output (and thus greater output is better), and multiply by 100,000.

For most DMUs in our data, we use data in Reilly et al. [3]. For the DMUs associated with Hurricanes Maria and Katrina, we gather these data from Fischbach et al. [33] and Energy [34], respectively.

3.3.2 Input Variables

Labor and capital are the key inputs to restore electrical power after a hurricane. This information can be proxied by the labor and capital costs incurred. Or more specifically, the costs of wire, pole, and transformer replacement costs; the cost of clean-up and restoration crews; and the cost of additional support personnel.

For most DMUs in our data, we use data in Reilly et al. [3]. For the DMUs associated with Hurricanes Maria and Katrina, we gather these data from [33] and from a variety of websites [35–38], respectively.

3.3.3 Environmental Variables

A previous study proposed that post-hurricane power restoration is primarily dependent on a set of hurricane severity measures and the size of the service area, namely wind speed and number of customers [3]. We include these, as well as peak number of customers without power as a proportion of all customers to account for the severity of the hurricane to a system. We gather these data for Hurricane Maria by noting there is one utility on Puerto Rico and overlaying this with preliminary National Institute of Standards and Technology calculations of peak wind gusts [39]. We gather these for Hurricane Katrina from overlaying the Commission [40] maps with the HAZUS peak wind gusts from FEMA [41].

Table 1 Summary of data used in the analysis

Storm	Utility	Total customer-days without power[a]	Restoration cost (in 2017 millions of dollars)	Service area (sq. miles) with winds		Percent of service area with winds		Peak number of customers without power	
				≥75 mph	≥100 mph	≥75 mph	≥100 mph	Number	Percent
Frances	A	7,040,000	410.0	6044	2944	0.24	0.12	2,800,000	0.651
Jeanne	A	3,160,000	417.8	4079	1445	0.16	0.06	1,740,000	0.405
Charley	A	2,650,000	327.0	2562	761	0.10	0.03	870,000	0.202
Isabel	B	7,320,000	170.6	14,007	345	0.47	0.01	1,690,000	0.845
Floyd	B	430,000	26.4	2351	0	0.05	0.00	520,000	0.260
Frances	C	2,710,000	168.7	2981	0	0.14	0.00	830,000	0.553
Jeanne	C	1,660,000	108.9	3731	0	0.18	0.00	720,000	0.480
Charley	C	2,260,000	197.2	2020	835	0.10	0.04	500,000	0.333
Ivan	D	1,990,000	124.5	8774	1115	0.20	0.02	830,000	0.593
Katrina	D	1,410,000	94.1	3420	0	0.08	0.00	630,000	0.450
Dennis	D	260,000	43.9	3236	161	0.07	0.00	240,000	0.171
Isabel	E	2,220,000	106.6	0	0	0.00	0.00	650,000	0.565
Irene	E	1,150,000	88.2	0	0	0.00	0.00	480,000	0.384
Sandy	E	260,000	40.5	0	0	0.00	0.00	210,000	0.168
Isabel	F	1,560,000	93.3	2	0	0.00	0.00	370,000	0.493
Isabel	G	330,000	17.9	3477	0	0.10	0.00	320,000	0.246
Charley	G	100,000	16.9	0	0	0.00	0.00	110,000	0.085
Jeanne	H	810,000	44.3	2490	0	0.62	0.00	290,000	0.450
Frances	H	460,000	32.8	1672	0	0.42	0.00	270,000	0.419

Fran	I	330,000	27.2	1123	0	0.05	0.00	260,000	0.113
Isabel	I	130,000	7.8	0	0	0.00	0.00	130,000	0.057
Ike	J	2,610,000	656.9	11,413	4043	0.60	0.21	382,202	1.000
Ike	K	13,120,000	853.8	1945	0	0.39	0.00	2,150,000	0.958
Irene	L	440,000	27.2	0	0	0.00	0.00	160,000	0.492
Sandy	L	620,000	42.7	0	0	0.00	0.00	187,000	0.575
Irene	M	970,000	85.0	0	0	0.00	0.00	410,000	0.373
Sandy	M	3,740,000	672.6	371	0	0.10	0.00	1,000,000	0.909
Maria	N	99,900,000	1700.0	3424	3081	1.00	0.90	1,569,796	1.000
Katrina	O	4,221,000	863.1	7600	4750	0.40	0.25	421,447	0.644
Katrina	P	317,000	349.4	3500	0	0.25	0.00	100,227	0.293
Katrina	Q	1,006,000	226.7	1400	1400	0.10	0.10	80,810	0.314
Katrina	R	6,737,000	356.4	350	350	1.00	1.00	215,163	1.000
Katrina	S	177,000	56.5	660	0	0.33	0.00	56,056	0.657

ª Until 90% of customers

4   Results

4.1   Stage 1 Super-Efficient Values

The preliminary efficiency values of electrical power restoration after moderate to severe hurricanes are calculated using the output-oriented, super-efficiency DEA method allowing for variable returns to scale. The super-efficiency DEA is bounded at zero; there is no upper bound. A higher score indicates greater efficiency. Ranking efficiency from least to most efficient in Table 2, utility companies have an average efficiency of 0.474 when we do not account for the environment. This ranges from 0.276 for the restoration efforts in Puerto Rico after Hurricane Maria to 1.379 for a utility company restoring electrical power after Hurricane Isabel.

4.2   Stage 2

In Stage 2, we regress the first-stage efficiency values on the non-discretionary inputs. The best model contained the log of the service area with peak wind gusts greater than 75 mph and the peak number of customers without power as a proportion of all customers. Because we can still have zero efficiency scores in "super-efficiency DEA", we run a Tobit regression; that said, results of the first stage do not contain zero. Test results of the Tobit regression model are statistically significant, where the likelihood ratio LR χ^2 (4) = 15.932 and the Prob> χ^2 = 0.000 (see Table 3). The findings indicate that the size of hurricane to a utility company in terms of the area facing peak winds over 75 mph is a useful factor in explaining efficiency. The smaller the service area with peak winds over 75 mph, the greater the efficiency in restoring power to customers, all else equal. Similarly, as the proportion of a utility's customers without power increases, the less efficient a utility company can be.

4.3   Stage 3 Final Super-Efficiency Values

The extent of the environmental indicators on multiple DMUs differ, and thus, utility companies facing more severe hurricanes will exhibit greater efficiency with an improved environment. Accordingly, we rank the DMUs in terms of the environmental severity each DMU faced with the worst DMU in terms of the non-discretionary inputs being ranked 1 and the least severe ranked 32. In the final estimation of efficiency, the DMU facing the most severe environment in the dataset is only compared to itself, which in this case is Ike for Utility J. Indeed, this is the same result as Reilly et al. [3], even though we find different variables have a statistically significant relationship with the first-stage super-efficiency value. This

Table 2 First-stage DEA results, least to most efficient

Storm	Utility	Year	First-stage efficiency
Maria	N	2017	0.276
Katrina	O	2005	0.303
Ike	K	2008	0.304
Sandy	M	2012	0.315
Ike	J	2008	0.316
Jeanne	A	2004	0.340
Frances	A	2004	0.341
Katrina	Q	2005	0.349
Katrina	P	2005	0.350
Charley	A	2004	0.354
Katrina	R	2005	0.378
Charley	C	2004	0.388
Isabel	B	2003	0.399
Frances	C	2004	0.400
Ivan	D	2004	0.425
Jeanne	C	2004	0.437
Isabel	E	2003	0.439
Katrina	D	2005	0.451
Isabel	F	2003	0.452
Irene	E	2011	0.458
Irene	M	2011	0.462
Katrina	S	2005	0.509
Jeanne	H	2004	0.541
Dennis	D	2005	0.542
Sandy	L	2012	0.547
Sandy	E	2012	0.554
Frances	H	2004	0.588
Fran	I	1996	0.621
Irene	L	2011	0.621
Floyd	B	1999	0.627
Isabel	G	2003	0.711
Isabel	I	2003	1.379
Unweighted mean			0.474

Table 3 Results of Tobit regression in Stage 2

Explanatory variables	Coefficient	SE	P value
Log service area with wind 75+	−0.016*	0.008	0.043
Peak number of customers (% of all customers)	−0.356**	0.106	0.001
Constant	0.751**	0.067	0.000

$**p < 0.01$; $*p < 0.05$. $N = 32$
LR χ^2 (4) = 15.932; Prob> χ^2 = 0.000

suggests it is a relatively robust finding that this utility-storm faced one of the worst environments in the data. The DMU facing the least severe environment is ranked 32, Isabel for Utility I, and is therefore compared to all other DMUs.

Efficiency value results are shown in Table 4. We discover that the average efficiency value increases from 0.474 to 0.836, whereas the lowest efficiency value of the sample increases from 0.276 to 0.491. The sole utility-storm for Hurricane Maria, receiving a first-stage value of 0.276, increased to 0.872 when controlling

Table 4 Final efficiency values

Second- stage Rank	Storm	Utility	Year	Third-stage super-efficiency	Third-stage traditional efficiency
8	Frances	A	2004	0.863	0.863
15	Jeanne	A	2004	0.628	0.628
23	Charley	A	2004	0.499	0.499
5	Isabel	B	2003	1.055	1.000
20	Floyd	B	1999	0.883	0.883
11	Frances	C	2004	0.787	0.787
12	Jeanne	C	2004	0.860	0.860
17	Charley	C	2004	0.660	0.660
9	Ivan	D	2004	–	1.000
13	Katrina	D	2005	0.888	0.888
25	Dennis	D	2005	0.879	0.879
24	Isabel	E	2003	0.618	0.618
29	Irene	E	2011	0.645	0.645
31	Sandy	E	2012	0.898	0.898
26	Isabel	F	2003	0.637	0.637
19	Isabel	G	2003	1.274	1.000
14	Jeanne	H	2004	1.064	1.000
16	Frances	H	2004	1.112	1.000
28	Fran	I	1996	0.874	0.874
32	Isabel	I	2003	–	1.000
1	Ike	J	2008	–	1.000
4	Ike	K	2008	0.804	0.804
22	Sandy	L	2012	0.769	0.769
27	Irene	L	2011	0.874	0.874
6	Sandy	M	2012	0.924	0.924
30	Irene	M	2011	0.650	0.650
2	Maria	N	2017	0.872	0.872
7	Katrina	O	2005	0.867	0.867
18	Katrina	P	2005	0.632	0.632
21	Katrina	Q	2005	0.643	0.643
3	Katrina	R	2005	0.893	0.893
10	Katrina	S	2005	–	1.000
	Mean			0.836	0.834

for the environment. Five utility-storms have efficiency levels greater than 1, which permits the relative assessment of even the most efficient utility-storm responses.

While we find no clear relationship by hurricane, results are largely consistent within utilities across hurricanes. That is, for the utility companies in our data responding to more than one hurricane, nearly all are consistently below full efficiency or consistently super-efficient. Utilities A, C, E, L, and M are less than fully efficient for all the post-hurricane power restorations in our data. Utility H was super-efficient (>1.0) in both the power restorations after Hurricanes Jeanne and Frances. Utilities D and I performance varied, such that they were super-efficient after one hurricane, but less than fully efficient in the other hurricanes in our data. When we examine results by hurricane, there is no clear pattern of utilities that perform well (i.e. super-efficient) or poorly.

Super-efficiency assessment can lead to infeasible solutions. For this reason, we also conducted a 'traditional' DEA in the third stage. We see that all the utility-storms with missing super-efficiency values have values of full 'traditional' efficiency (=1.0). Cook et al. [42] propose a technique for modifying infeasible super-efficiency solutions.

We tested whether there was any relationship with time and find no statistically significant effect on third-stage efficiency values, suggesting it is not the case that utility companies become more efficient in later years of the data.

This study finds that the average increase in efficiency when accounting for environment is 0.38. These findings conformably illustrate that most companies dealing with more severe hurricanes will demonstrate an efficiency improvement after incorporating hurricane severity factors that contribute to restoring electrical power.

5 Discussion

This paper builds off the framework developed in Reilly et al. [3], and makes three contributions to the literature on systematic evaluations of the performance of electric power companies in restoring service following hurricanes. First, we demonstrate the implications of using a super-efficiency approach. We find 25% (eight out of 32) of disasters have an efficiency value of 1.0. After running a super-efficiency approach, we can rank these disaster responses with values ranging from 1.05 to 1.28 (plus infeasible solutions). We find that the application of super-efficiency DEA for assessing electrical power restorations after a hurricane is a worthwhile approach given a significant proportion of restorations, approximately 1 in 4, were fully efficient and can be compared using this super-efficiency DEA method.

Second, this study includes an additional environmental variable that turns out to be statistically important for relating the severity of a hurricane to the relative efficiency of a restoration. Specifically, we include peak number of customers as a proportion of all customers in the system. This highlights the importance of not just the absolute number of customers needing power restored, but the number relative to the size of the system. The greater the proportion of customers without service, controlling for number of customers, the greater inefficiency observed.

Third, this study assesses more recent storms that are considered relatively severe and for which there was a great deal of controversy regarding the time in which power was restored, including Hurricanes Katrina and Maria. Indeed, first stage results that do not account for environment would suggest the restorations were highly inefficient, with a value of 0.276 for utility N-Maria and 0.303 to 0.509 for Hurricane Katrina responses. However, after considering the severity of the hurricanes, results indicate the utility N-Maria restoration was weakly efficient with a value of 0.87. Of the storms in more favorable environmental contexts, the utility N-Maria restoration is relatively more efficient than 50% of utility-storm restorations (15 out of 30); it is the median value restoration as opposed to the least efficient in the data set. Similarly, for the Katrina restorations, efficiency values improved 41% to 216%. This shows the importance of assessing restorations accounting for the environmental context.

This study finds post-hurricane power restoration performance is largely consistent within utilities across hurricanes, but we find no clear relationship by hurricane. These results are similar to findings in Reilly et al. [3]. The implication of this result is that efficiency is indeed within the control of firms, not just regional infrastructure and environmental characteristics. This study suggests models should consider more of these firm-specific characteristics, such as their supply-chains or logistics, relationships with suppliers and prices, and labor resources, to better predict outages and duration of outages. While the private sector owns and operates most of the U.S. energy infrastructure, they are subject to mandatory federal reliability standards to ensure operational reliability. Exploring the more efficient DMUs in this study may help the private and public sector identify best practice technology and improve best standards.

One question that may arise from the findings for these more recent hurricane restorations is whether they could have been more efficient. There are two ways to answer this question using DEA. One way of interpreting DEA results is that by adopting best practice technology, utility N-Maria had the scope of producing 1.147 times (i.e. 1/0.872) as much output, in terms of restoring power to more customers faster, from the same level of inputs. Another way DEA results can be used to answer this question is to compare the restorations performed in environments that would be considered worse. According to the environmental variables used in the analysis of the second stage, one DMU, Utility J after Hurricane Ike, operated in a more challenging context than Utility N after Hurricane Maria. Compared to Utility J-Ike, the Utility N-Maria restoration was less efficient.

Acknowledgements We thank Kurt Klein and Clark Gardner for downloading data necessary for the work. We thank Jeff Wenger, Isaac Opper, and M. Granger Morgan for helpful discussions on relevant topics. The authors would like to thank two anonymous referees for thoughtful suggestions that improved this work.

References

1. Hines, P., Apt, J. & Talukdar, S. 2009. Large Blackouts In North America: Historical Trends And Policy Implications. *Energy Policy,* 37, 5249-5259.
2. Campbell, R. 2012. Weather-Related Power Outages And Electric System Resiliency. Congressional Research Service.
3. Reilly, A. C., Davidson, R. A., Nozick, L. K., Chen, T. & Guikema, S. D. 2016. Using Data Envelopment Analysis To Evaluate The Performance Of Post-Hurricane Electric Power Restoration Activities. *Reliability Engineering & System Safety,* 152, 197-204.
4. Gorman, M. & Ruggiero, J. 2008. Evaluating Us State Police Performance Using Data Envelopment Analysis. *International Journal Of Production Economics,* 113, 1031-1037.
5. Fried, H. O., Lovell, C. A. K., Schmidt, S. S. & Yaisawarng, S. 2002. Accounting For Environmental Effects And Statistical Noise In Data Envelopment Analysis. *Journal Of Productivity Analysis,* 17, 157-174.
6. National Academies Of Sciences, E. & Medicine 2017. *Enhancing The Resilience Of The Nation's Electricity System,* Washington, Dc, The National Academies Press.
7. Vaiman, Bell, Chen, Chowdhury, Dobson, Hines, Papic, Miller & Zhang 2012. Risk Assessment Of Cascading Outages: Methodologies And Challenges. *Ieee Transactions On Power Systems,* 27, 631-641.
8. Castillo, A. 2014. Risk Analysis And Management In Power Outage And Restoration: A Literature Survey. *Electric Power Systems Research,* 107, 9-15.
9. Wang, Y., Chen, C., Wang, J. & Baldick, R. 2016. Research On Resilience Of Power Systems Under Natural Disasters—A Review. *Ieee Transactions On Power Systems,* 31, 1604-1613.
10. Liu, Y. & Zhong, J. Risk Assessment Of Power Systems Under Extreme Weather Conditions — A Review. 2017 Ieee Manchester Powertech, 18-22 June 2017 2017. 1-6.
11. Arab, A., Khodaei, A., Han, Z. & Khator, S. K. 2015. Proactive Recovery Of Electric Power Assets For Resiliency Enhancement. *Ieee Access,* 3, 99-109.
12. Arif, A., Ma, S. & Wang, Z. Online Decomposed Optimal Outage Management After Natural Disasters. 2017 Ieee Power & Energy Society General Meeting, 16-20 July 2017 2017. 1-5.
13. Arif, A., Ma, S., Wang, Z., Wang, J., Ryan, S. M. & Chen, C. 2018. Optimizing Service Restoration In Distribution Systems With Uncertain Repair Time And Demand. *Ieee Transactions On Power Systems,* 33, 6828-6838.
14. Barabadi, A. & Ayele, Y. Z. 2018. Post-Disaster Infrastructure Recovery: Prediction Of Recovery Rate Using Historical Data. *Reliability Engineering & System Safety,* 169, 209-223.
15. Guikema, S. D., Nateghi, R., Quiring, S. M., Staid, A., Reilly, A. C. & Gao, M. 2014. Predicting Hurricane Power Outages To Support Storm Response Planning. *Ieee Access,* 2, 1364-1373.
16. Liu, H., Davidson, R. A. & Apanasovich, T. V. 2007. Statistical Forecasting Of Electric Power Restoration Times In Hurricanes And Ice Storms. *Ieee Transactions On Power Systems,* 22, 2270-2279.
17. Reed, D. A. 2008. Electric Utility Distribution Analysis For Extreme Winds. *Journal Of Wind Engineering And Industrial Aerodynamics,* 96, 123-140.
18. Nateghi, R. 2018. Multi-Dimensional Infrastructure Resilience Modeling: An Application To Hurricane-Prone Electric Power Distribution Systems. *Ieee Access,* 6, 13478-13489.
19. Nateghi, R., Guikema, S. D. & Quiring, S. M. 2014. Forecasting Hurricane-Induced Power Outage Durations. *Natural Hazards,* 74, 1795-1811.
20. Mukherjee, S., Nateghi, R. & Hastak, M. 2018. A Multi-Hazard Approach To Assess Severe Weather-Induced Major Power Outage Risks In The U.S. *Reliability Engineering & System Safety,* 175, 283-305.
21. Wanik, D. W., Anagnostou, E. N., Hartman, B. M., Frediani, M. E. B. & Astitha, M. 2015. Storm Outage Modeling For An Electric Distribution Network In Northeastern Usa. *Natural Hazards,* 79, 1359-1384.
22. Emrouznejad, A. & Yang, G.-L. 2018. A Survey And Analysis Of The First 40 Years Of Scholarly Literature In Dea: 1978–2016. *Socio-Economic Planning Sciences,* 61, 4-8.

23. Liu, J. S., Lu, L. Y., Lu, W.-M. & Lin, B. J. 2013. A Survey Of Dea Applications. *Omega*, 41, 893-902.
24. Sueyoshi, T., Yuan, Y. & Goto, M. 2017. A Literature Study For Dea Applied To Energy And Environment. *Energy Economics*, 62, 104-124.
25. Banker, R. D., Charnes, A. & Cooper, W. W. 1984. Some Models For Estimating Technical And Scale Inefficiencies In Data Envelopment Analysis. *Management Science*, 30, 1078-1092.
26. Cooper, W. W., Seiford, L. M. & Tone, K. 2000. Data Envelopment Analysis. *Handbook On Data Envelopment Analysis, 1st Ed.; Cooper, Ww, Seiford, Lm, Zhu, J., Eds*, 1-40.
27. Nahra, T. A., Mendez, D. & Alexander, J. A. 2009. Employing Super-Efficiency Analysis As An Alternative To Dea: An Application In Outpatient Substance Abuse Treatment. *European Journal Of Operational Research*, 196, 1097-1106.
28. Andersen, P. & Petersen, N. C. 1993. A Procedure For Ranking Efficient Units In Data Envelopment Analysis. *Management Science*, 39, 1261-1264.
29. Shepherd, R. W. 2015. *Theory Of Cost And Production Functions*, Princeton University Press.
30. Ruggiero, J. 1996. On The Measurement Of Technical Efficiency In The Public Sector. *European Journal Of Operational Research*, 90, 553-565.
31. Ruggiero, J. 1998. Non-Discretionary Inputs In Data Envelopment Analysis. *European Journal Of Operational Research*, 111, 461-469.
32. Muñiz, M., Paradi, J., Ruggiero, J. & Yang, Z. 2006. Evaluating Alternative Dea Models Used To Control For Non-Discretionary Inputs. *Computers & Operations Research*, 33, 1173-1183.
33. Fischbach, J., May, L., Whipkey, K., Shelton, S., Vaughan, C., Tierney, D., Leuschner, K., Meredith, L. & Peterson, H. 2020. After Hurricane Maria: Predisaster Conditions, Hurricane Damage, and Recovery Needs in Puerto Rico. Santa Monica, CA: RAND Corporation, Report #RR-2595-DHS.
34. Energy, D. O. 2020. *Iser - Hurricane Katrina* [Online]. Available: Https://Www.Oe.Netl.Doe.Gov/Hurricanes_Emer/Katrina.Aspx Last Accessed February 27, 2020. [Accessed].
35. Cleco. 2018. *Cleco Katrina/Rita Hurricane Recovery Funding Llc* [Online]. Available: Https://Www.Cleco.Com/-/Cleco-Katrina-Rita-Hurricane-Recovery-Funding-Llc Last Accessed February 27, 2020 [Accessed].
36. Dixie. 2020. *Dixie Electric Power Association Hurricane Katrina Remembrance Day* [Online]. Available: Https://Www.Cooperative.Com/Content/Public/2016-Spotlight-Entries/16-Event/Best%20event%20class%202%20gold%20dixie.Pdf Last Accessed February 27, 2020. [Accessed].
37. Newsroom, E. 2018. *Entergy Louisiana's Katrina And Rita Restoration Costs Are Paid In Full* [Online]. Available: Https://Www.Entergynewsroom.Com/News/Entergy-Louisianakatrina-Rita-Restoration-Costs-Are-Paid-Full/ Last Accessed February 27, 2020 [Accessed].
38. Nola.Com. 2015. *Entergy Learns Katrina Lessons, But Damage Prevention Still In Question* [Online]. Available: Https://Www.Nola.Com/News/Article_Bb5f05b6-701b-5a6b-Adfd-42f64f30b5b7. [Accessed].
39. Nist 2020. Ncst Technical Investigation Of Hurricane Maria's Impacts On Puerto Rico: Preliminary Project Plan For Characterization Of Hazards. National Institute Of Standards And Technology.
40. Commission, L. P. S. 2020. *Louisiana Electric Utilities Area Maps* [Online]. Available: Http://Www.Lpsc.Louisiana.Gov/Maps_Electric_Distribution_Areas.Aspx Last Access February 27, 2020 [Accessed].
41. Fema 2020. Hazus Success Stories.
42. Cook, W. D., Liang, L., Zha, Y. & Zhu, J. 2009. A Modified Super-Efficiency Dea Model For Infeasibility. *The Journal Of The Operational Research Society*, 60, 276-281.

Transparency Tools for Fairness in AI (Luskin)

Mingliang Chen, Aria Shahverdi, Sarah Anderson, Se Yong Park, Justin Zhang, Dana Dachman-Soled, Kristin Lauter, and Min Wu

Abstract We propose new tools for policy-makers to use when assessing and correcting fairness and bias in AI algorithms. The three tools are:

- A new definition of fairness called "controlled fairness" with respect to choices of protected features and filters. The definition provides a simple test of fairness of an algorithm with respect to a dataset. This notion of fairness is suitable in cases where fairness is prioritized over accuracy, such as in cases where there is no "ground truth" data, only data labeled with past decisions (which may have been biased).
- Algorithms for retraining a given classifier to achieve "controlled fairness" with respect to a choice of features and filters. Two algorithms are presented, implemented and tested. These algorithms require training two different models

Our suite of tools is christened "Luskin," as this project began at the UCLA Meyer and Renee Luskin Conference Center.

Mingliang Chen and Aria Shahverdi authors contributed equally to this work.

Dana Dachman-Soled and Aria Shahverdi were supported in part by NSF grants #CNS-1933033, #CNS-1840893, #CNS-1453045 (CAREER), by a research partnership award from Cisco and by financial assistance award 70NANB15H328 from the U.S. Department of Commerce, National Institute of Standards and Technology.

M. Chen · A. Shahverdi (✉) · D. Dachman-Soled · M. Wu
University of Maryland, College Park, MD, USA
e-mail: mchen126@umd.edu; ariash@umd.edu; danadach@umd.edu; minwu@umd.edu

S. Anderson
University of St. Thomas, St. Paul, MN, USA
e-mail: ande1298@stthomas.edu

S. Y. Park · J. Zhang
University of Maryland, College Park, MD, USA

Montgomery Blair High School, Silver Spring, MD, USA

K. Lauter
Microsoft Research, Redmond, WA, USA
e-mail: klauter@microsoft.com

in two stages. We experiment with combinations of various types of models for the first and second stage and report on which combinations perform best in terms of fairness and accuracy.

- Algorithms for adjusting model parameters to achieve a notion of fairness called "classification parity." This notion of fairness is suitable in cases where accuracy is prioritized. Two algorithms are presented, one which assumes that protected features are accessible to the model during testing, and one which assumes protected features are not accessible during testing.

We evaluate our tools on three different publicly available datasets. We find that the tools are useful for understanding various dimensions of bias, and that in practice the algorithms are effective in starkly reducing a given observed bias when tested on new data.

Keywords Controlled fairness · Machine learning · Policy

1 Introduction

Machine Learning (ML) is a set of valuable mathematical and algorithmic tools, which use existing data (known as "training data") to learn a pattern, and predict future outcomes on new data based on that pattern. ML algorithms are the foundation of a revolution in Artificial Intelligence (AI), which is replacing humans with machines. In business, public policy and health care, decision-makers increasingly rely on the output of a trained ML classifier to make important and life-altering decisions such as whether to grant an individual a loan, parole, or admission to a college or a hospital for treatment. In recent years it has become clear that, as we use AI or ML algorithms to make predictions and decisions, we perpetuate bias that is inherent in the training data.

In this paper, we propose new tools for policy-makers to use when assessing and correcting fairness and bias in AI algorithms. Datasets typically come in the form of databases with rows corresponding to individual people or events, and columns corresponding to the features or attributes of the person or event. Some features are considered sensitive and may be "protected," such as race, gender, or age. Typically we are interested in preventing discrimination or bias based on "protected" features, at least in part due to the fact that it is illegal. Other features may be either highly correlated with protected features or may be relevant from a common sense point of view with the decision to be made by the classifier. By "protected classes" we refer to the different groups arising from the various settings of the protected feature.

The notion of fairness is a rather complex one and there are multiple aspects of fairness and/or perceived fairness related to machine learning (cf. [8, 10, 12, 13, 15]). This work is not intended as a survey and therefore we discuss only the fairness definitions most closely related to the current work. We focus on two angles, and argue that each is appropriate in different real-life settings:

Controlled Fairness In cases where a machine learning algorithm is trained on data labeled with prior *decisions*, as opposed to an objective "ground truth," our main concern is *parity*, across protected classes. For example, in a stop-question-frisk law enforcement setting, a classifier deciding whether or not a person should be frisked, is trained on past stop-question-frisk data. But whether a person was frisked or not in the past, depends on a (potentially biased) decision. Thus, this is a setting where achieving accuracy with respect to past decisions may not be desirable. Another setting in which *parity* is our main concern (taking precedence over accuracy) is a setting in which an organization is compelled to bear the risk of an incorrect decision, in order to improve societal welfare. For example, a bank may use a classifier to predict whether a person will default on a loan, and then use this information to determine whether to approve the loan. In this case, our main concern may be that the classifier achieves *parity* across protected classes, as long as the bank's overall risk does not significantly increase.

Our fairness definition requires that controlling for certain "unprotected attributes" such as *type of crime* in the law-enforcement example, or *education* level in the loan example, the classifier's $0/1$ output rate with respect to a given dataset is approximately the same across the protected classes. We believe this definition will be useful to practitioners as it provides an explicit test for fairness of a specific algorithm with respect to a specific dataset. Of course, we still want to achieve the best accuracy possible with respect to the prior decisions (since the point of training the machine learning model in the first place is to obtain an algorithm that emulates human decisions), while ensuring that the fairness conditions are met.

By controlling for "unprotected features," we are taking the point of view that it can make sense to "filter" datasets based on "unprotected features" which seem relevant to making a good decision. Filtering simply selects various rows which satisfy certain conditions on the entries in specified columns. This can be viewed as a controlled experiment, which allows one to determine the bias stemming from the protected feature, as opposed to other confounding factors. After filtering, we propose to test for fairness with respect to protected features by checking whether ratios of outcomes are approximately the same across the protected classes.

Accuracy-Based Fairness In cases where accuracy with respect to a ground truth is prioritized, the above definition may not be the right choice. In this case, the notion of fairness we consider ensures that the true positive rate (TPR) and false positive rate (FPR) are as close as possible across the protected classes, while overall accuracy remains high. We enforce this equivalent learning performance across all protected classes to the extent possible, sacrificing little on the high performance of the model on the majority protected class. This coincides with the *equalized odds* [12] and *classification parity* [5] notion that have previously been considered in the literature. An appropriate setting for applying this notion of fairness is in predicting recidivism, where there is arguably a more objective "ground truth." Specifically, taking as an example the parole decisions, the machine learning algorithm is not trained on data labeled by the decisions themselves, but rather on data labeled according to whether or not a person who was released was subsequently re-arrested. Furthermore, the risk of making an incorrect decision is extremely high and cannot simply be absorbed as a loss, as in the case of loan defaults. It has been documented (and our own experiments support this) that in the recidivism case, when machine learning algorithms are trained on available data, non-whites have both higher TPR and FPR than whites. This implies that a larger fraction of non-whites than whites are predicted to re-commit a crime, when, in fact, they do not go on to re-commit a crime within 2 years. Equalizing the TPR and FPR for whites and non-whites (to the extent possible)—and simultaneously maintaining the overall prediction accuracy of the classifier—results in improved fairness outcomes across the protected classes.

1.1 The Toolkit

The three tools we introduce in this work are:

- A new definition of *controlled fairness* with respect to choices of protected features and filters. This definition provides a simple test of fairness of an algorithm with respect to a dataset.
- Algorithms for retraining a classifier to achieve "controlled fairness" with respect to a choice of features and filters. Two algorithms are presented, implemented and tested. These algorithms require training two different models in two stages. We experiment with combinations of various types of models for the first and second stage and report on which combinations perform best in terms of fairness and accuracy.
- Algorithms for adjusting model parameters to achieve "classification parity." Two algorithms are presented, one which assumes that protected features are accessible to the model during testing, and one which assumes protected features are not accessible during testing.

We also implement and evaluate our tools on the Stop-Question-Frisk dataset from NYPD, the Adult Income dataset from the US Census and the COMPAS recidivism dataset. The Stop, Question and Frisk dataset is a public record of an individual who has been stopped by NYPD, and it contains detailed information about the incident, such as time of the stop, location of the stop, etc. The Adult Income dataset was taken from the 1994 Census Database, and each row has information about an individual such as marital status, education, etc. The COMPAS recidivism dataset records individuals' basic information and their recidivism within 2 years. The basic information includes race, age, crime history, etc.

We find that the tools are useful for understanding various dimensions of bias, and that in practice the tools are effective for eliminating a given observed bias when tested on new data.

1.2   Prior Fairness Definitions and Our Contributions

Establishing a formal definition of fairness that captures our intuitive notions and ensures desirable outcomes in practice, is itself a difficult research problem. The legal and machine learning literature has proposed different and conflicting definitions of fairness [8, 10, 12, 13, 15]. We begin with an overview of some definitions of fairness from the literature and their relation to the notions studied in this work.

We first discuss the relationship of our new notion of "controlled fairness" to the prior notion of "statistical parity," originally introduced in [8]. We then discuss our contributions with respect to definitions, algorithms and implementations in this regime.

Next, we discuss the prior notion of "classification parity" [5] and its relationship to our work on accuracy-based fairness. We discuss our contributions with respect to definitions, algorithms and implementations in this regime.

Statistical Parity The notion of statistical parity stems from the legal notion of "disparate impact" [2]. Other names in the literature for the same or similar notion include *demographic parity* [3, 23] and *group fairness* [22]. This notion essentially requires that the outcome of a classifier is equalized across the protected classes. For example, it may require that the percentage of female and male applicants accepted to a college is approximately the same.

More formally, the *independence* notion that underlies the definition of statistical parity (see e.g. [1]) requires that, in the case of binary classification:

$$\mathbb{P}\{R = 1 \mid A = a\} = \mathbb{P}\{R = 1 \mid A = b\}. \tag{1}$$

Here, A corresponds to the protected feature, which can be set to value a or b. R is the random variable corresponding to the output of the classifier on an example sampled from some distribution $\mathcal{D}$. Thus, the left side of Eq. (1) corresponds to the probability that a classifier outputs 1 on an example sampled from $\mathcal{D}$, *conditioned on the protected feature of the example being set to value a*. Similarly, the right side of Eq. (1) corresponds to the probability that a classifier outputs 1 on an example sampled from $\mathcal{D}$, *conditioned on the protected feature of the example being set to value b*.

Assuming rows of a database are sampled as i.i.d. random variables from distribution $\mathcal{D}$, the left hand side probability in (1) can be approximated as the **ratio of** *the number of rows in the database for which the classifier outputs 1 and the protected feature is set to a* **to** *the number of rows in the database for which the protected feature is set to a*. Analogously, the right hand side probability in (1) can be approximated as the **ratio of** *the number of rows for which the classifier outputs 1 and the protected feature is set to b* **to** *the number of rows for which the protected feature is set to b*. These empirical ratios are the basis of our controlled fairness definition (which will be introduced formally in Sect. 3).

The independence-based notions discussed above have been criticized in the machine learning literature, mainly due to the fact that they inherently sacrifice accuracy, since the *true classification* may itself be deemed "unfair" under these definitions. Our notion of "controlled fairness" is a refinement of these notions (as it allows, in addition, filtering on unprotected features), and indeed may preclude achieving perfect accuracy. Due to this limitation, our notion should only be applied in situations where "fairness" is prioritized over "accuracy." For example, in some situations such as college admissions, there is no "ground truth" to measure accuracy against, only data about previous admission decisions, which may themselves have been biased. Therefore, in such settings, our goal should not be solely to achieve optimal accuracy with respect to past decisions.

Other objections to these notions include the fact that it may not always be desirable to equalize the outcomes across the protected classes. In our above example, if 20% of the female applicants have GPA of at least 3.5 and SAT scores of at least 1500, while only 10% of the male applicants have GPA of at least 3.5 and SAT scores of at least 1500, then one may argue that it is "fair" for a larger percentage of the female applicants to be accepted than the male.

Our notion of fairness remedies exactly this situation, by allowing "controls" or "filters" to be placed on features that are considered "unprotected." In the above example, a "filter" selects the set of rows satisfying the condition that the GPA is at least 3.5 and SAT score at least 1500. Then, among those selected rows, we require that the percentage of accepted females and males is approximately the same.

In summary, our work on *controlled fairness* makes the following contributions and/or distinctions beyond the notion of statistical parity and other notions previously considered in the literature:

1. We introduce a definition of "controlled fairness" that allows one to identify a "filtering" condition on unprotected features that permits one to enforce parity of outcomes across subgroups of the protected classes. (Unlike other definitions in [11] which enforces an intersection on protected features.) This addresses some of the objections to the statistical parity notion, since it does not mandate naïve equalization across the protected classes, but allows for subtleties in how fairness across protected classes is evaluated.
2. Our proposed notion prioritizes equality in predictions, and does not consider a data-generating mechanism or a ground-truth data. We emphasize that the focus is on stating whether a classifier achieves the controlled fairness notion with respect to a given dataset. Thus, we are not mainly concerned with accuracy in this setting and consider it especially appropriate to apply this notion in cases where there is no "ground truth" data, but only data on prior decisions (which may have been biased).
3. We propose two algorithms to retrain a classifier in settings when the originally trained classifier is not fair with respect to our fairness notion.
4. We test our algorithms on real-world data and report the outcomes. Our algorithms require training two different models in two stages. We experiment

with combinations of various types of models for the first and second stage and report on which combinations perform best in terms of fairness and accuracy.

Comparison with ϵ-conditional Parity We note that our proposed definition is perhaps most similar to the notion of ϵ-*conditional parity*, introduced by Ritov et al. [16]. The main difference between our notions is that our notion is concrete: It specifies whether a specific classifier does or does not achieve "controlled fairness" with respect to a specific dataset. On the other hand, the notion of [16] defines the fairness of a classifier with respect to its output distribution on instances drawn from various conditional distributions. This makes it less useful as a tool for checking and enforcing fairness with respect to a particular dataset.

Classification Parity Other notions of fairness considered in the machine learning literature include *classification parity, calibration, calibration within groups* and *balance for the positive/negative class* [5, 15]. These notions of fairness prioritize accuracy with respect to a "ground truth," since a classifier that outputs the true labels will always satisfy the definition. In our work, we recognize the subtleties in the application of the various fairness notions (i.e. not every fairness notion is suitable for application in all situations). We therefore consider applying these accuracy-based notions, specifically, the notion of *classification parity* (as opposed to our previously introduced notion of controlled fairness), in the case that accuracy is prioritized. *Classification parity* ensures that certain common measures of predictive performance are (approximately) equal across the protected groups. Under this definition, a classifier that predicts recidivism, for example, would be required to produce similar false positive rates for white and black parole applicants. In particular, we will focus on obtaining a classifier whose learning performances across the protected classes are as similar as possible. In other words, we should not be able to distinguish which class of instances is being tested just from the learning performance statistics.

In summary, our work on *classification parity* has the following contributions:

1. We formulate a *weak* and *strong* condition on classification parity and propose two approaches for achieving improved fairness in classifiers.
2. The first algorithm trains a single classifier and then tunes the decision thresholds for each of the protected classes to achieve classification parity. This methodology can be shown to improve fairness for each of the protected classes by equalizing the true positive/false positive rates across classes. However, in order to know which thresholds to apply on an input instance, the classifier must know the values of the protected features.
3. The second algorithm is applicable in the case that it is not legally or socially acceptable to use different classifiers for each of the protected classes, or in the case in which the protected features are simply not known to the classifier. In the algorithm, a single classifier is trained by incorporating a trade-off between accuracy and fairness. The fairness is quantitatively measured by the *equalized distribution* (a notion we introduce in Sect. 4) of positive/negative instances across the protected classes.

1.3   Additional Related Work

Prior Work on Achieving Statistical Parity and Removing Discrimination   Prior work has suggested entirely different algorithms to achieve similar goals as the goal of this work, which is to obtain classifiers that achieve the controlled fairness definition. For example, the recent work of Wang et al. [18], focuses on the statistical parity notion and suggests to search for a perturbed distribution, which they call a "counterfactual distribution" on which disparity of the classifier across the classes is minimized. Then, for each input to the classifier, they perform a pre-processing step that modifies the features of the input example according to the counterfactual distribution and then run the original classifier on the modified input. The recent work of Udeshi et al. [17] takes as input a potentially unfair classifier and searches the input space to find "discrimatory examples"—two inputs that are highly similar, differ with respect to the protected feature, and are classified differently. Then, using "corrected" labels on these discriminatory examples, the original classifier can be retrained to improve its performance.

In contrast to the work of Wang et al. [18], our approach does not require a preprocessing step to be applied to the test input by the end-user. Instead, the final classifier can run as before on a test input. This allows for simplicity and backwards compatibility for the end-user, and would be a more socially acceptable solution, since no overt modification of inputs is performed by the end-user (the only modifications occur during training).

In contrast to both the works of Udeshi et al. [17] and Wang et al. [18], our approach is conceptually simple, and can be performed by running a standard training algorithm as-is to generate the final model that is outputted. There is no additional search step that receives the description of the model and must find either the so-called "counterfactual distribution," or "discriminatory examples," both of which require additional complex and non-standard algorithms. This makes our approach more suitable as part of a toolkit for policy-makers.

Prior Work on Achieving Accuracy-Based Fairness   A large number of researches have investigated different variations of accuracy-based fairness and improved the fairness achieved by classifiers. These fair algorithms mainly fall into two categories: those that have prior knowledge of the protected feature in test stage [9, 12] and those that do not have prior knowldge [4, 19–21].

Hardt et al. [12] focused on tuning the classifier to satisfy fairness constraints after the classifier has been trained. The basic idea is to find proper thresholds on receiver operating characteristic (ROC) curve where the classifier meets classification parity. Dwork et al. [9] proposed an algorithm to select a set of classifiers out of a larger set of classifiers according to a joint loss function that balances accuracy and fairness. The advantage is that two works can be applied to any given classifiers since no retraining or modification is needed for the classifiers. However, these methods require access to the protected feature in test stage. Hardt et al. [12] did not balance the accuracy and fairness during the process of threshold searching, and also does not guarantee to obtain the optimal solution. Dwork et al. [9] needed large overhead in training multiple classifiers for all the data groups. In contrast, our classification parity based method only requires one trained classifier and the post-processing can provide the unique optimal solution balancing accuracy and fairness.

Learning a new representation for the data is another approach in fairness learning [4, 21], removing the information correlated to the protected feature and preserving the information of data as much as possible. For example, Zemel et al. [21] introduced a mutual information concept in learning the fair representation of the data. The advantage of this approach is that it can be applied before the classifier learning stage and no prior knowledge of the protected feature is required during test-time. The weakness is the lack of the flexibility in accuracy and fairness tradeoff.

A direct way to tradeoff accuracy and fairness is to incorporate a constraint or a regularization of fairness term into the training optimization objective of the classifier. The fairness term is described as classification parity, such as demographic parity [20], equalized odds [12] and predictive rate parity [19]. Note that these methods also do not require access to the protected feature in the test stage. In our work, we propose a new definition of fairness that ensures a strong version of classification parity, which we call *equalized distribution*. Compared with the prior art on classification parity, such as equalized odds [12], equalized distribution is a stronger condition on classification parity, requiring that the classifier achieves equivalent performance statistics (e.g., TPR and FPR) among all the groups, independent of the decision threshold settings of the classifier.

1.4 Organization

The rest of this paper is organized as follows. In Sect. 2 we introduce the notation we use throughout the paper. In Sect. 3, we begin by presenting our new definition of *controlled fairness*. We then present two algorithms to obtain classifiers that achieve the controlled fairness notion (See Sects. 3.1 and 3.2). Next, we describe the datasets used to evaluate the performance of our algorithms in Sect. 3.3. In Sect. 3.4 we discuss the reasoning behind our high-level implementation choices and in Sect. 3.5 we evaluate the performance—fairness and accuracy—of the proposed algorithms. In Sect. 4 we propose two algorithms to achieve classification parity in classifier training, and we show the experimental results on the fairness improvements with the proposed methods.

2 Preliminaries

Notation Dataset D is represented as a matrix in $\mathbb{R}^{s \times n}$. It has features $\{f_1, \ldots, f_n\}$ (corresponding to columns $1, \ldots, n$), rows $D^1, \ldots, D^s$ and columns $D_1, \ldots, D_n$. The j-th entry of the i-th row is denoted by $D[i][j]$. We denote by $\#\{D\}$ the number of rows in D (similarly for a set S, we denote by $\#S$ the cardinality of the set). In practice, some features are considered protected, while others are considered unprotected. We denote by $F^P \subseteq \{f_1, \ldots, f_n\}$ the set of protected features and by $F^N \subseteq \{f_1, \ldots, f_n\}$ the set of unprotected features.

Database Operators In the database literature, the output of a SQL "Select" query with condition *cond* on dataset D is represented as the output of operator $\sigma_{cond}(D)$. A SQL "Select" query with condition *cond* on dataset D returns a dataset that consists of all rows of D satisfying condition *cond*, i.e. if the entries in certain columns satisfy the condition set for those columns. Thus, $|\sigma_{cond}(D)|$ means the number of rows of D satisfying condition *cond*.

Classifiers and Risk Assignment Algorithms A Classifier is an algorithm which takes as input a row in dataset and returns a discrete set of values as its output (in our case we assume binary classifier–yes/no), representing the predicted class for the input. Similarly, a risk assignment algorithm takes a input a row in dataset and returns a real number (interpreted as a probability of being a yes/no instance) between 0 and 1. A risk assignment algorithm is denoted by A and the corresponding classifier is represented by A^C. Specifically, classifier A^C is obtained from risk assignment algorithm A as follows,

$$A^C(D^i) = \begin{cases} 1, & A(D^i) \geq \text{threshold} \\ 0, & \text{otherwise} \end{cases}$$

Syntax for Learning Algorithms A learning algorithm $\mathcal{M}$ is an algorithm that takes as input a *labeled dataset* $D^+ \in \mathbb{R}^{s \times (n+1)}$ and outputs a binary classifier C. The labeled dataset D^+ consists of rows $D^{1,+}, \ldots, D^{s,+}$ and columns $D_1^+, \ldots, D_{n+1}^+$. Each row $D^{i,+} \in \mathbb{R}^n \times \{0, 1\}$ is called an *example*. An example $D^{i,+} = D^i || b^i$ consists of a feature vector $D^i \in \mathbb{R}^n$ that corresponds to a setting of the features $\{f_1, \ldots, f_n\}$ and a label b^i.

3 Controlled Fairness

In this section, we introduce our new fairness definition, present algorithms for obtaining classifiers that achieve this notion and evaluate the performance of our algorithms on real datasets.

We introduce the following definition of controlled fairness for a binary classification setting:

Definition 1 (Controlled Fairness) Let $D \in \mathbb{R}^{s \times n}$ be a dataset, let $cond_N$ be a set of conditions on some unprotected features in F^N and let $cond_P$ be a condition on a protected feature in F^P. Let $\neg cond_P$ denote the negation of the condition. Let

$D^+ \in \mathbb{R}^{s \times (n+1)}$ denote the *labeled* dataset obtained by running (binary) classifier C on dataset D.

Define the datasets DNP, ¬DNP to be the output of the SQL "Select" queries

$$\text{DNP} := \sigma_{cond_N, cond_P}(D^+) \quad \neg\text{DNP} := \sigma_{cond_N, \neg cond_P}(D^+)$$

We require that #{DNP} > 0 and #{¬DNP} > 0.

Define the ratio of a dataset as follows:

$$\text{ratio}(D) = \frac{\#\{k \mid D[k][n+1] = 1\}}{\#\{D\}}$$

We say that C is a fair classifier with respect to $D, cond_P, cond_N$ if:

$$\text{ratio}(\text{DNP}) \approx \text{ratio}(\neg\text{DNP})$$

We next introduce two algorithms for achieving controlled fairness. We then implement and experimentally validate our algorithms on multiple datasets.

Let $D_{(1)}$ be a dataset and let $A_{(1)}$ be a risk assignment algorithm trained on this dataset. $A_{(1)}^C$ is the corresponding classifier, which we assume does not achieve the controlled fairness notion as it was defined in Definition 1.

Specifically, given a new dataset $D_{(2)}$, we label the dataset with the output of the classifier $A_{(1)}^C$ to obtain $D_{(2)}^+$. Since classifier $A_{(1)}^C$ does not achieve the controlled fairness notion (see Definition 1) with respect to $D_{(2)}$ (if it did, we would be done) we have that

$$\text{ratio}(\text{DNP}_{(2)}) \not\approx \text{ratio}(\neg\text{DNP}_{(2)}),$$

where

$$\text{DNP}_{(2)} := \sigma_{cond_N, cond_P}(D_{(2)}^+) \quad \neg\text{DNP}_{(2)} := \sigma_{cond_N, \neg cond_P}(D_{(2)}^+)$$

Our goal is to remove the bias from $A_{(1)}^C$, even though we do not have access to unbiased training data. In the following two subsections we present two algorithms to solve this problem. Specifically, we would like to leverage $A_{(1)}^C$ to train a new classifier $C_{(2)}$ that remains accurate but is unbiased. The idea of the following algorithms are to use $A_{(1)}^C$ to selectively relabel the dataset $D_{(2)}^+$ such that the biased is removed. We denote the relabeled dataset by $D_{(3)}^+$ and we refer to $D_{(3)}^+$ as a *synthetic* dataset, since the labels of $D_{(3)}^+$ do not correspond to either the original labels or the labels produced by a classifier. Specifically, we will consider two ways of constructing $D_{(3)}^+$ such that the following is satisfied:

$$\text{ratio}(\text{DNP}_{(3)}) \approx \text{ratio}(\neg\text{DNP}_{(3)}),$$

where

$$\mathsf{DNP}_{(3)} := \sigma_{cond_N, cond_P}(D_{(3)}^+) \quad \neg\mathsf{DNP}_{(3)} := \sigma_{cond_N, \neg cond_P}(D_{(3)}^+)$$

We then train a new classifier $C_{(2)}$ on the synthetic dataset. We expect $C_{(2)}$ to achieve the controlled fairness definition with respect a newly sampled dataset $D_{(4)}$ which has never been seen by the classifier, while accuracy with respect to the true labels remains high. We will validate these expectations with our experimental results in Sect. 3.5. In the following, we describe two algorithms for generating the synthetic labeled dataset $D_{(3)}^+$.

3.1   Algorithm 1: Synthetic Data via Selective Risk Adjustment

In this section we present our first proposed algorithm. The idea of this algorithm is to adjust the risk values associated with one of the protected classes, compute new labels based on the adjusted risk values and output the resulting database as $D_{(3)}^+$. Recall that $A_{(1)}$ is trained on dataset $D_{(1)}$ and that labeled dataset $D_{(2)}^+$ is obtained by applying classifier $A_{(1)}^C$ on a new dataset $D_{(2)}$. Specifically, the labeled dataset is computed as follows:

$$D_{(2)}^+[i][n+1] = \begin{cases} 1, & \text{if } A_{(1)}(D_{(2)}^i) \geq \mathsf{threshold} \\ 0, & \text{otherwise} \end{cases}$$

For the case of logistic regression the **threshold** in the above equation is usually set at 0.5.

Since we assumed classifier $A_{(1)}^C$ does not already achieve the controlled fairness definition, we have WLOG that $\mathsf{ratio}(\mathsf{DNP}_{(2)}) \geq \mathsf{ratio}(\neg\mathsf{DNP}_{(2)})$. In particular:

$$\frac{\#\{k \mid A_{(1)}(\mathsf{DNP}_{(2)}^k) \geq \mathsf{threshold}\}}{\#\{\mathsf{DNP}_{(2)}\}} \geq \frac{\#\{k \mid A_{(1)}(\neg\mathsf{DNP}_{(2)}^k) \geq \mathsf{threshold}\}}{\#\{\neg\mathsf{DNP}_{(2)}\}} \tag{2}$$

Let

$$\alpha := \frac{\#\{\mathsf{DNP}_{(2)}\} \cdot \#\{k \mid A_{(1)}(\neg\mathsf{DNP}_{(2)}^k) \geq \mathsf{threshold}\}}{\#\{\neg\mathsf{DNP}_{(2)}\}}$$

To obtain synthetic dataset $D_{(3)}^+$, we first compute Δ such that the following holds:

$$\#\{k \mid \left(A_{(1)}(\mathsf{DNP}_{(2)}^k) - \Delta\right) \geq \mathsf{threshold}\} \approx \alpha \tag{3}$$

Fig. 1 Note that the quantity $\#\{k \mid A_{(1)}(\text{DNP}^k_{(2)}) \geq$ threshold$'\}$ is represented by the area under the curve and to the right of the vertical line passing through (threshold$'$, 0). Therefore, we set the value of threshold$'$ so that the area of the marked region is equal to α. Then Δ is set to threshold$'$ − threshold

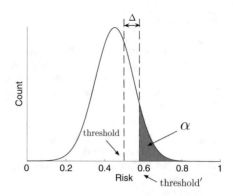

In order to compute Δ we sort $\text{DNP}_{(2)}$ according to the risk value outputted by $A_{(1)}$ on each entry and find the maximal value of threshold$'$ such that $\#\{k \mid A_{(1)}(\text{DNP}^k_{(2)}) \geq \text{threshold}'\} \approx \alpha$. Then Δ can be computed as $\Delta = \text{threshold}' - \text{threshold}$. Figure 1 shows an example of the distribution of $A_{(1)}(\text{DNP}_{(2)})$ and pictorially represents the method of finding Δ.

$\text{DNP}_{(3)}$ is equivalent to $\text{DNP}_{(2)}$, except for the final column (the $(n + 1)$-st column), which corresponds to the labels. The labels of $\text{DNP}_{(3)}$ are defined as follows:

$$\text{DNP}_{(3)}[i][n + 1] = \begin{cases} 1, & \text{if } A_{(1)}(\text{DNP}^i_{(2)}) - \Delta \geq \text{threshold} \\ 0, & \text{otherwise} \end{cases}$$

It is straightforward to see that the following property is satisfied

$$\frac{\#\{k \mid \text{DNP}_{(3)}[k][n + 1] = 1\}}{\#\{\text{DNP}_{(3)}\}} \approx \frac{\#\{k \mid \neg\text{DNP}_{(2)}[k][n + 1] = 1\}}{\#\{\neg\text{DNP}_{(2)}\}}.$$

$D^+_{(3)}$ is then defined to be a concatenation of the following datasets.

$$D^+_{(3)} := \text{DNP}_{(3)} \mid \neg\text{DNP}_{(2)} \mid \sigma_{\neg cond_N}(D^+_{(2)})$$

The synthetic dataset $D^+_{(3)}$ will be used to train a new classifier $C_{(2)}$. Algorithm 1 shows pseudocode for this algorithm. The FILTER function takes a dataset D, and returns the rows which satisfies whatever conditions $cond$ is passed to it as input. The CHECKFAIR function takes two dataset and the label column and check whether the fairness notion, as it is introduced in Definition 1, is satisfied. The CHECKFAIR function returns 1 if fairness notion is achieved and returns 0, otherwise.

Algorithm 1 Synthetic data via selective risk adjustment

1: **procedure** MAIN($D, A, colnum, rownum, th, cond_P, cond_N$)
2:     Call RISKASSIGNMENT($D, A, colnum, rownum$)
3:     Call INITIALIZELABELS($D, colnum, rownum, th$)
4:     Call SYNTHETICGEN($D, colnum, rownum, th, cond_P, cond_N$)
5:     $C \leftarrow$ LEARN($D, colnum + 1, rownum$)
6:     Output C
7: **end procedure**

8: **procedure** RISKASSIGNMENT($D, A, colnum, rownum$)
9:     **for** $i = 1$ to $rownum$ **do**
10:         Set $D[i][colnum + 2] = A(D[i][1], \ldots, D[i][colnum])$
11:     **end for**
12: **end procedure**

13: **procedure** INITIALIZELABELS($D, colnum, rownum, th$)
14:     **for** $i = 1$ to $rownum$ **do**
15:         **if** $D[i][colnum + 2] \geq th$ **then**
16:             Set $D[i][colnum + 1] = 1$
17:         **else**
18:             Set $D[i][colnum + 1] = 0$
19:         **end if**
20:     **end for**
21: **end procedure**

22: **procedure** SYNTHETICGEN($D, colnum, rownum, th, cond_P, cond_N$)
23:     Set DNP $=$FILTER($D, cond_N, cond_P$)
24:     Set ¬DNP $=$FILTER($D, cond_N, \neg cond_P$)
25:     Set ¬DN $=$FILTER($D, \neg cond_N$)
26:     **if** CHECKFAIR(DNP, ¬DNP, $colnum + 1$) $= 0$ **then**
27:         sum $= 0$
28:         **for** $i = 1$ to #{¬DNP} **do**
29:             **if** ¬DNP$[i][colnum + 2] \geq th$ **then**
30:                 sum $=$ sum $+ 1$
31:             **end if**
32:         **end for**
33:         $\alpha = \frac{\#\{\text{DNP}\} \cdot \text{sum}}{\#\{\neg\text{DNP}\}}$
34:         Sort DNP according to $colnum + 2$ from largest to smallest
35:         $th' =$ DNP[round(α)][$colnum + 2$]
36:         $\Delta = th' - th$
37:         **for** $i = 1$ to #{DNP} **do**
38:             DNP$[i][colnum + 2] =$ DNP$[i][colnum + 2] - \Delta$
39:         **end for**
40:         Call INITIALIZELABELS(DNP, $colnum$, #{DNP}, th)
41:         $D :=$ Concatenate(DNP, ¬DNP, ¬DN)
42:     **end if**
43: **end procedure**

3.2 Algorithm 2: Synthetic Data via Risk Based Flipping

In this section, we present our second proposed algorithm. The idea of this algorithm is to preserve the original labels of the datapoints in $D_{(2)}$ (as opposed to using $A_{(1)}^C$ to fully relabel the dataset) and flip only the minimal number of datapoints within one of the protected classes to construct $D_{(3)}^+$ that satisfies

$$\text{ratio}(\text{DNP}_{(3)}) \approx \text{ratio}(\neg\text{DNP}_{(3)}).$$

$A_{(1)}$ will be utilized to provide a risk score that helps decide which datapoints should be flipped.

Recall that as in the previous sections classifier $A_{(1)}^C$ is trained on dataset $D_{(1)}$. In this section, we assume a new *labeled* dataset $D_{(2)}^+$ is given, labeled with the true labels. If labeled dataset $D_{(2)}^+$ is not biased, then we can trivially set $D_{(3)}^+ := D_{(2)}^+$. Therefore, we assume WLOG that the following holds

$$\frac{\#\{k \mid \text{DNP}_{(2)}[k][n+1] = 1\}}{\#\{\text{DNP}_{(2)}\}} \geq \frac{\#\{k \mid \neg\text{DNP}_{(2)}[k][n+1] = 1\}}{\#\{\neg\text{DNP}_{(2)}\}}.$$

Similar to the previous algorithm we fix the right hand side:

$$\alpha := \frac{\#\{\text{DNP}_{(2)}\} \cdot \#\{k \mid \neg\text{DNP}_{(2)}[k][n+1] = 1\}}{\#\{\neg\text{DNP}_{(2)}\}}$$

The above indicates that we should construct $DNP_{(3)}$ such that it has α number of rows with label 1. Since there are currently $\#\{k \mid \text{DNP}_{(2)}[k][n+1] = 1\}$ rows in $\text{DNP}_{(2)}$ with label 1, we must flip the labels for $\beta = \alpha - \#\{k \mid \text{DNP}_{(2)}[k][n+1] = 1\}$ number of rows from 1 to 0. To do so, we first divide $\text{DNP}_{(2)}$ into disjoint datasets based on its labels and flip enough samples only from the dataset with label of 1. Let $cond_L$ be the condition that selects the datapoints of $\text{DNP}_{(2)}$ with label of 1 and $\neg cond_L$ be the condition that selects datapoints with label of 0. Then we construct the following two datasets from $\text{DNP}_{(2)}$,

$$\text{DNPL}_{(2)} := \sigma_{cond_N, cond_P, cond_L}(D_{(2)}^+) \quad \neg\text{DNPL}_{(2)} := \sigma_{cond_N, cond_P, \neg cond_L}(D_{(2)}^+)$$

Then, we sort $\text{DNPL}_{(2)}$ according to the score of each data point, as assigned by $A_{(1)}$, from smallest to largest. $\text{DNPL}_{(3)}$ is equivalent to (the sorted version of) $\text{DNPL}_{(2)}$, except for the final column (the $(n+1)$-st column), which corresponds to the labels. The labels of $\text{DNPL}_{(3)}$ are defined as follows:

$$\text{DNPL}_{(3)}[i][n+1] = \begin{cases} 0, & \text{if } i \leq \beta \\ 1, & \text{otherwise} \end{cases}$$

We then define $\mathsf{DNP}_{(3)}$ as follows: $\mathsf{DNP}_{(3)} := \mathsf{DNPL}_{(3)}|\neg\mathsf{DNPL}_{(2)}$ and it is straightforward to see that the following property is satisfied

$$\frac{\#\{k \mid \mathsf{DNP}_{(3)}[k][n+1] = 1\}}{\#\{\mathsf{DNP}_{(3)}\}} \approx \frac{\#\{k \mid \neg\mathsf{DNP}_{(2)}[k][n+1] = 1\}}{\#\{\neg\mathsf{DNP}_{(2)}\}}.$$

Similar to the previous algorithm, we define the synthetic dataset $D_{(3)}^{+}$ to be a concatenation of the following datasets.

$$D_{(3)}^{+} := \mathsf{DNP}_{(3)} \quad | \quad \neg\mathsf{DNP}_{(2)} \quad | \quad \sigma_{\neg cond_N}(D_{(2)}^{+})$$

The new dataset $D_{(3)}^{+}$ will be used to train a new classifier $C_{(2)}$. Algorithm 2 shows pseudocode for this algorithm. The FILTER function takes a dataset D, and returns the rows which satisfies whatever conditions *cond* is passed to it as input. The CHECKFAIR function takes two dataset and the label column and check whether the fairness notion, as it is introduced in Definition 1, is satisfied. The CHECKFAIR function returns 1 if fairness notion is achieved and returns 0, otherwise.

3.3   Choice of Dataset

Stop, Question and Frisk Dataset The *Stop, Question and Frisk* dataset is a publicly available dataset that consists of information collected by New York Police Department officers since 2003 [6]. We selected the data from year 2012 as it had a sufficient number of entries for training purposes. The original dataset from 2012 had 532,911 rows and 112 columns and after cleaning 477,840 rows remained. Additionally, we selected 33 relevant columns, based on the description which is provided with the dataset. Each row of the dataset represents an individual who has been stopped by an officer and includes detailed information about the incident such as time of stop, reason for stop, crime they are suspected of, etc.

Figure 2 shows the majority of the people in the dataset (i.e. a majority of people who were stopped by an officer) are non-white. In addition, a significantly higher proportion of non-whites are frisked compared to whites. Figure 3 shows that when we filter on the type of crime an individual is suspected of, e.g. assault vs. non-assault, Non-Whites are still more likely to be frisked. Finally, Fig. 4 presents the percentage of frisked individuals within each race Fig. 3. It can be seen that White individuals are less likely to be frisked, even when suspected of the same type of crime (assault).

Adult Income Dataset The *Adult Income* dataset was taken from the 1994 Census Database and contains 15 columns of demographic information collected by the census. The dataset contains 48,842 entries. The task for this dataset was to predict whether a given individual earned more than $50K per year. We chose this dataset since such a classifier can then be used, for example, to decide whether an individual

---

Algorithm 2 Synthetic data via risk based flipping

1: **procedure** MAIN($D, A, colnum, rownum, th, cond_P, cond_N, cond_L$)
2:    Call RISKASSIGNMENT($D, A, colnum, rownum$)
3:    Call SYNTHETICGEN($D, colnum, rownum, th, cond_P, cond_N, cond_L$)
4:    $C \leftarrow$ LEARN($D, colnum + 1, rownum$)
5:    Output C
6: **end procedure**

7: **procedure** RISKASSIGNMENT($D, A, colnum, rownum$)
8:    **for** $i = 1$ to $rownum$ **do**
9:       Set $D[i][colnum + 2] = A(D[i][1], \ldots, D[i][colnum])$
10:    **end for**
11: **end procedure**

12: **procedure** SYNTHETICGEN($D, colnum, rownum, th, cond_P, cond_N, cond_L$)
13:    Set DNP =FILTER($D, cond_N, cond_P$)
14:    Set DNPL =FILTER($D, cond_N, cond_P, cond_L$)
15:    Set ¬DNPL =FILTER($D, cond_N, cond_P, ¬cond_L$)
16:    Set ¬DNP =FILTER($D, cond_N, ¬cond_P$)
17:    Set ¬DN =FILTER($D, ¬cond_N$)
18:    **if** CHECKFAIR(DNP, ¬DNP, $colnum + 1$) = 0 **then**
19:       sum = 0
20:       **for** $i = 1$ to #{¬DNP} **do**
21:          **if** ¬DNP$[i][colnum + 1] == 1$ **then**
22:             sum = sum + 1
23:          **end if**
24:       **end for**
25:       $\alpha = \frac{\#\{DNP\} \cdot sum}{\#\{¬DNP\}}$
26:       $\beta = \alpha - \#\{DNPL\}$
27:       Sort DNPL according to $colnum + 2$ from smallest to largest
28:       **for** $i = 1$ to β **do**
29:          DNPL$[i][colnum + 1] = 0$
30:       **end for**
31:       $D :=$ Concatenate(DNPL, ¬DNPL, ¬DNP, ¬DN)
32:    **end if**
33: **end procedure**

---

gets approved for a loan. Three columns were dropped from the dataset when training the models. Native country was dropped because it did not contain a lot of information, as the vast majority of individuals were from the United States, and using the column as a training feature would have added a high degree of dimensionality. Education level was also dropped because the same information was encoded in a separate numeric column. Finally, fnlwgt—the statistical weight assigned to each individual by the census—was dropped because it had extremely low correlation with the target column.

Figure 5 shows that the majority of individuals in the dataset are white, and a greater proportion of White people make more than $50K per year than Non-whites. Figure 6 shows the same plot filtered on education level (only for the subset of the dataset with more than 10 years of education). Figure 7 represent the percentage for

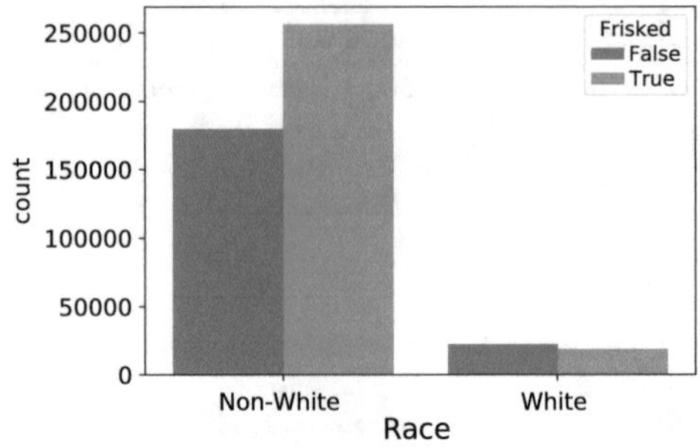

Fig. 2 Number of people
who got frisked in each race

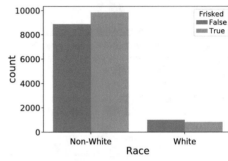

Fig. 3 Number of people
who got frisked conditioned
on being suspected of
committing an assault

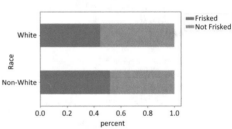

Fig. 4 Comparison of
frisked ratio between races
conditioned on being
suspected of committing an
assault

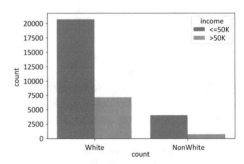

Fig. 5 Number of people by
race and income level

each race for the Fig. 6. It can be seen that a higher percentage of White individuals
versus Non-White make more than $50K, even when controlling for education level.

Fig. 6 Number of people by race and income level conditioned on having more than 10 years of education

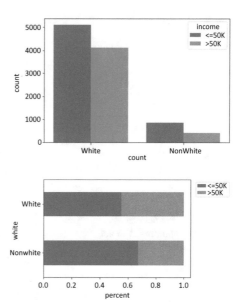

Fig. 7 Comparison of ratio of income levels conditioned on having more than 10 years of education

3.4 Type of Models for Our Algorithms

Algorithms 1 and 2 both require two trained models, denoted $A_{(1)}$ and $C_{(2)}$. In this section, we will consider the choice of the *type* of model used for $A_{(1)}$ and $C_{(2)}$. Recall that $A_{(1)}$ is a risk assignment algorithm. This means we have to select a type of model which outputs continuous values representing the probability of each output class. Luckily, two of the most popular types of models "Logistic Regression (LR)" and "Multi-layer Perceptron (MLP)," have this property. While $C_{(2)}$ is only required to be a classifier, we still consider the same two types of models for it. Therefore, there are 4 different choices for the types for $A_{(1)}/C_{(2)}$: LR/LR, LR/MLP, MLP/LR, MLP/MLP. As our experimental data will show, some combinations of types perform significantly better than others, both in the case of Algorithms 1 and 2. In the following we will provide a high-level explanation for why both Algorithms 1 and 2 perform better when $C_{(2)}$ is of type MLP.

For illustrative purposes, we use the *Adult Income Dataset* to construct a simplified model in which only two features—age and hours per week—are considered. In this case, age is the protected feature and hours per week is the unprotected feature. We will observe that the synthetic dataset constructed by Algorithm 1 and 2 is difficult for a LR-type model to learn.

Algorithm 1 Recall that in this case we find an optimal value of Δ and use it to perturb the dataset $D_{(2)}^{+}$, which is labeled with the output of $A_{(1)}^{C}$. Figure 8 shows the prediction of the first classifier, i.e. $A_{(1)}^{C}$ on dataset $D_{(2)}$. The decision boundary for that classifier is also represented in the same figure. Figure 9 shows the new labels produced after applying Δ to subset of dataset $D_{(2)}$ to generate a synthetic

Fig. 8 $A_{(1)}^C$ predictions on dataset $D_{(2)}$

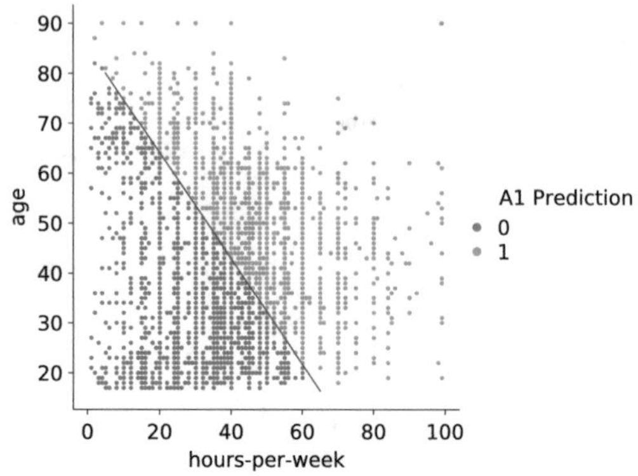

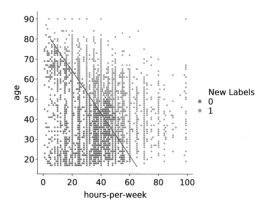

Fig. 9 $A_{(1)}^C$ Predictions adjusted by applying Δ to get synthetic dataset $D_{(3)}^+$

dataset $D_{(3)}^+$. Note that applying Δ creates non linear decision boundaries (observe the blue triangular region that appears), resulting in a dataset that cannot be learned to high accuracy with logistic regression.

Algorithm 2 Recall that in this case we keep the original labels of dataset $D_{(2)}^+$ and flip only a minimal number of labels (using $A_{(1)}$ to decide which labels to flip). Figure 10 shows the decision boundary of $A_{(1)}^C$ as well as original labels of $D_{(2)}^+$. Figure 11 shows the new labels generated by flipping enough sample points in the subset of dataset $D_{(2)}^+$ to generate synthetic dataset $D_{(3)}^+$. Similar to the previous algorithm, flipping these labels creates non linear decision boundaries (a blue triangular region again appears, although fainter than before), resulting in a dataset that cannot be learned to high accuracy with logistic regression.

It is therefore crucial for the second classifier to be able to find these non-linear boundaries as well. As supported by the experimental data, Multi-layer Perceptron, which potentially has non-linear decision boundaries, is therefore a good choice for the type of the second classifier $C_{(2)}$.

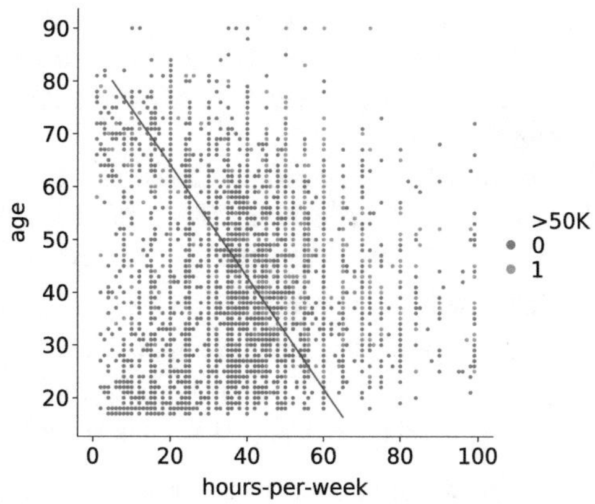

Fig. 10 $A^C_{(1)}$ decision boundary with original labels of $D_{(2)}$

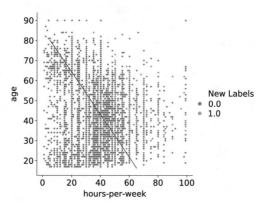

Fig. 11 $A^C_{(1)}$ decision boundary and flipping labels of $D_{(2)}$ to get synthetic dataset $D^+_{(3)}$

3.5 Experimental Result

In this section we present the result of both of the algorithms explained earlier. In each of the tables there are protected features shown in the first column. For both of the datasets we chose to use "race" as a protected feature, with "W" representing White and "NW" representing non-White. The unprotected features and corresponding conditions, shown by "Filter" in the table, are different for each dataset. For the *Stop, Question and Frisk* dataset, the unprotected feature is *type of crime* and the corresponding condition is whether an individual is suspected of committing an assault, represented by "A" or not suspected of assault, represented by "NA." For the *Adult Income* dataset, the unprotected feature is *education level* and the corresponding condition is whether an individual has more than 10 years of education, represented by "E" or not, represented by "UE" in the tables. We also report the accuracy of each trained model, represented by "Acc" in the tables, in terms of the Area under Curve (AUC) of each model. The accuracy is measured with respect to the original labels. In each case we report the ratio, as it is introduced in Definition 1, of each dataset given the conditions on its columns. The final goal is to have a classifier which achieves fairness notion as it was introduced in Definition 1. Specifically, ratio should be approximately the same across the protected classes. For the choice of model type, as explained in Sect. 3.4, we consider 4 different combinations.

To measure the performance of our algorithms, we divide our dataset into 3 batches of size $40\%, 40\%, 20\%$. The first 40% of dataset is $D_{(1)}$ and is used to construct the first classifier $A_{(1)}^C$. The second 40% of dataset is $D_{(2)}$ and is used to first check the fairness notion for $A_{(1)}^C$ and construct the new dataset, i.e. $D_{(3)}^+$. The last 20% is $D_{(4)}$ which is being used for testing the fairness of both $A_{(1)}^C$ and $C_{(2)}$. All the numbers we report correspond to the performance of the classifier with respect to the dataset $D_{(4)}$, which has never been seen by the classifier.

Stop, Question and Frisk Dataset Table 1 shows the performance—fairness and accuracy—for Algorithms 1 and 2 on the *Stop, Question and Frisk Dataset* when the first model, $A_{(1)}$, is of type *Logistic Regression*. The first 2 columns are obtained from the original labels of the data. They show that when we filter on "Assault", the ratio of individuals who got frisked is 45% and 52% for White and Non-White, respectively—a gap of 7%. The next 2 columns are obtained from the labels generated by $A_{(1)}^C$ on the dataset. In this case, when we filter on "Assault", the ratio of frisked individuals is 25% and 38% for White and Non-White, respectively—a gap of 13%. By selecting the second model, $C_{(2)}$, to be of type *Logistic Regression*, the ratios become 20% and 27% for Algorithm 1 and 19% and 25% for Algorithm 2. Note that while the difference in the ratios is not significantly reduced (as expected when using a second model of type LR). Once we choose MLP as the type of the second model, $C_{(2)}$, the ratios become 25% and 27% for Algorithm 1 and 42% and 46% for Algorithm 2, reducing the gap to 2% and 4%, respectively. Note that the gap is comparable for the two algorithms, Algorithm 2 is preferable, since the ratios of 42% and 46% are much closer to the original ratios of 45% among Whites (the ratio in the original data), which we are trying to match. Indeed, the accuracy of the second model produced by Algorithm 1 is about 2% lower than the accuracy of the second model produced by Algorithm 2. Note that some drop in accuracy is expected, since we are measuring accuracy with respect to the original labels, and our goal is to ensure fairness (which is not satisfied by the original labels).

Table 2 shows the performance—fairness and accuracy—for Algorithms 1 and 2 on the *Stop, Question and Frisk Dataset* when the first model, $A_{(1)}$, is of type *Multi-layer Perceptron*. The first 2 columns are obtained from the original labels of the data. They show that when we filter on "Assault", the ratio of individuals who got frisked is 45% and 52% for White and Non-White, respectively—a gap of 7%. The next 2 columns are obtained from the labels generated by $A_{(1)}^C$ on the dataset. In this case, when we filter on "Assault", the ratio of frisked individuals is 25% and 38% for White and Non-White, respectively—a gap of 13%. By selecting the second model, $C_{(2)}$, to be of type *Logistic Regression*, the ratios become 23% and 31% for Algorithm 1 and 19% and 24% for Algorithm 2. Note that the difference in the ratios is not significantly reduced (as expected when using a second model of type LR). Once we choose MLP as the type of the second model, $C_{(2)}$, the ratios

Table 1 Performance of Algorithms 1 and 2 on Stop, Question and Frisk Dataset when the first model is of type *Logistic Regression*

	Original data		$A_{(1)}^C$ (LR)		$C_{(2)}$ (LR)		$C_{(2)}$ (MLP)	
Algorithm 1								
Acc			0.8186		0.8181		0.7972	
Filter	NA	A	NA	A	NA	A	NA	A
W	0.456	0.445	0.321	0.258	0.320	0.199	0.323	0.246
NW	0.588	0.520	0.525	0.380	0.525	0.271	0.525	0.271
Algorithm 2								
Acc			0.8186		0.8183		0.8160	
Filter	NA	A	NA	A	NA	A	NA	A
W	0.456	0.445	0.321	0.258	0.327	0.188	0.466	0.423
NW	0.588	0.520	0.525	0.380	0.528	0.247	0.604	0.458

Table 2 Performance of Algorithms 1 and 2 on Stop, Question and Frisk Dataset when the first model is of type *Multi-Layer Perceptron*

	Original data		$A_{(1)}^C$ (MLP)		$C_{(2)}$ (LR)		$C_{(2)}$ (MLP)	
Algorithm 1								
Acc			0.8167		0.8157		0.8168	
Filter	NA	A	NA	A	NA	A	NA	A
W	0.456	0.445	0.401	0.412	0.362	0.230	0.401	0.378
NW	0.588	0.520	0.567	0.502	0.536	0.305	0.566	0.396
Algorithm 2								
Acc			0.8167		0.8183		0.8172	
Filter	NA	A	NA	A	NA	A	NA	A
W	0.456	0.445	0.401	0.412	0.326	0.188	0.433	0.473
NW	0.588	0.520	0.567	0.502	0.527	0.242	0.625	0.476

become 38% and 40% for Algorithm 1 and 47% and 48% for Algorithm 2, reducing the gap to 2% and 1%, respectively. Note that while the gap and the overall accuracy is comparable for the two algorithms, Algorithm 2 is preferable, since the ratios of 47% and 48% are much closer to the original ratios of 45% among Whites (the ratio in the original data), which we are trying to match.

Adult Income Dataset Table 3 shows the performance—fairness and accuracy—for Algorithms 1 and 2 on the *Adult Income Dataset* when the first model, $A_{(1)}$, is of type *Logistic Regression*. The first 2 columns are obtained from the original labels of the data. They show that when we filter on "Education", the ratio of individuals who earn more than 50k is 44% and 36% for White and Non-White, respectively—a gap of 8%. The next 2 columns are obtained from the labels generated by $A_{(1)}^C$ on the dataset. In this case, when we filter on "Education", the ratio of individuals who earn more than 50k is 62% and 46% for White and Non-White, respectively—a gap of 16%. By selecting the second model, $C_{(2)}$, to be of type *Logistic Regression*, the ratios become 63% and 52% for Algorithm 1 and 61% and 53% for Algorithm 2.

Table 3 Performance of Algorithms 1 and 2 on Adult Income Dataset when the first model is of type *Logistic Regression*

	Original data		$A_{(1)}^C$ (LR)		$C_{(2)}$ (LR)		$C_{(2)}$ (MLP)	
Algorithm 1								
Acc			0.9005		0.8917		0.8998	
Filter	UE	E	UE	E	UE	E	UE	E
W	0.165	0.442	0.279	0.620	0.302	0.630	0.281	0.620
NW	0.088	0.358	0.143	0.463	0.195	0.515	0.160	0.578
Algorithm 2								
Acc			0.9005		0.8995		0.8984	
Filter	UE	E	UE	E	UE	E	UE	E
W	0.165	0.442	0.279	0.620	0.260	0.610	0.100	0.457
NW	0.088	0.358	0.143	0.463	0.200	0.526	0.049	0.414

Note that the difference in the ratios is only slightly reduced (as expected when using a second model of type LR). Once we choose MLP as the type of the second model, $C_{(2)}$, the ratios become 62% and 58% for Algorithm 1 and 46% and 41% for Algorithm 2, reducing the gap to 4% and 5%, respectively. Note that while the gap and accuracy is comparable for the two algorithms, Algorithm 2 is preferable, since the ratios of 46% and 41% are much closer to the original ratios of 44% among Whites (the ratio in the original data), which we are trying to match.

Table 4 shows the performance—fairness and accuracy—for Algorithms 1 and 2 on *Adult Income* when the first model, $A_{(1)}$, is of type *Multi-layer Perceptron*. Similarly, the **ratio** of individuals who earn more than 50*k* is 44% and 27% for White and Non-White, respectively. Using a second model of type *Logistic Regression* reduces the difference to 5%. Using a second model of type *Multi-layer Perceptron* reduces the difference even further to 2% for Algorithm 2. There is also no significant drop in terms of accuracy of the classifier.

Table 4 shows the performance—fairness and accuracy—for Algorithms 1 and 2 on the *Adult Income Dataset* when the first model, $A_{(1)}$, is of type *Multi-layer Perceptron*. The first 2 columns are obtained from the original labels of the data. They show that when we filter on "Assault", the **ratio** of individuals who earn more than 50*k* is 44% and 36% for White and Non-White, respectively—a gap of 8%. The next 2 columns are obtained from the labels generated by $A_{(1)}^C$ on the dataset. In this case, when we filter on "Education", the **ratio** of individuals who earn more than 50*k* is 44% and 28% for White and Non-White, respectively—a gap of 16%. By selecting the second model, $C_{(2)}$, to be of type *Logistic Regression*, the ratios become 55% and 50% for Algorithm 1 and 61% and 53% for Algorithm 2. Note that the difference in the ratios is only slightly reduced in the case of Algorithm 2 (as expected when using a second model of type LR). Once we choose MLP as the type of the second model, $C_{(2)}$, the ratios become 46% and 40% for Algorithm 1 and 44% and 41% for Algorithm 2, reducing the gap to 6% and 3%, respectively. Note that while the gap and the overall accuracy is comparable for the two algorithms,

Table 4 Performance of Algorithms 1 and 2 on Adult Income Dataset when the first model is of type *Multi-Layer Perceptron*

	Original data		$A_{(1)}^C$ (MLP)		$C_{(2)}$ (LR)		$C_{(2)}$ (MLP)	
Algorithm 1								
Acc			0.8994		0.8927		0.9036	
Filter	UE	E	UE	E	UE	E	UE	E
W	0.165	0.442	0.086	0.442	0.131	0.549	0.087	0.457
NW	0.088	0.358	0.071	0.276	0.118	0.500	0.083	0.403
Algorithm 2								
Acc			0.8994		0.8991		0.8982	
Filter	UE	E	UE	E	UE	E	UE	E
W	0.165	0.442	0.086	0.442	0.259	0.605	0.095	0.435
NW	0.088	0.358	0.071	0.276	0.203	0.526	0.047	0.410

Algorithm 2 is preferable, since the ratios of 44% and 41% are much closer to the original ratios of 44% among Whites (the ratio in the original data), which we are trying to match.

4   Fairness via Classification Parity

In this section, we investigate the fairness issue from statistical learning perspective. In Bayesian decision theory for binary classification, one sample s is classified into negative class ω_0 or positive ω_1 by one decision threshold, which is determined according to the posterior probability distribution. For example, given the distribution and the threshold shown in Fig. 12a, we have the relation of true positive (TP), true negative (TN), false positive (FP) and false negative (FN). Specially for the minimum error classifier, the decision threshold is set as Fig. 12b, where $P(\omega_0|s) = P(\omega_1|s)$. When the decision threshold is tuned, the true positive rate (TPR) and false positive rate (FPR) are altered accordingly.

In general, a binary classifier C can be considered as a mapping function of the input sample x to the posterior probability or score s, i.e., $C : x \rightarrow s$. Denote y and $\hat{y}$ are the corresponding label and classifier prediction of x, and $y, \hat{y} \in \{0, 1\}$. The classifier's output scores from the group of data formulates the score distribution shown in Fig. 12. The decision criterion is

$$\hat{y} = \begin{cases} 0, & s = C(x) < t \\ 1, & s = C(x) \geq t \end{cases} \tag{4}$$

where t is the decision threshold. When we tune the decision threshold for one trained classifier C, we can obtain different TPR and FPR according to the score distribution, which formulates C's ROC curve.

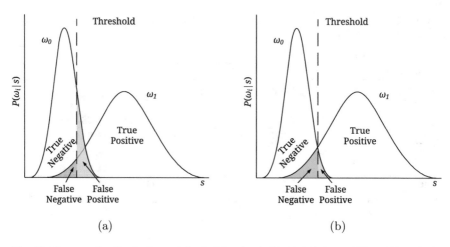

Fig. 12 The posterior distribution and the decision threshold. (**a**) Relations of TP and FP with the given threshold. (**b**) The threshold of the minimum error classifier

4.1 Classification Parity

Suppose the dataset D are separated into multiple completely exhaustive mutually exclusive groups $D_1, D_2, \ldots, D_K$ based on the protected attribute. For example, if one classifier has higher TPR and FPR in the group D_k than the other group $D_i, i \neq k$, the classifier is prone to provide positive inferences in D_k. We refer to the above condition as *positively biased* in D_k. Prior art of the fairness definition, such as demographic parity [20], equalized odds [12] and predictive rate parity [19], can alleviate the biased prediction in the classifier by equalizing the performance statistics of the classifier among all the groups. However, these classification parity conditions guarantees that the classifier only satisfies the parity condition at one specific threshold setting. If we modify the decision threshold, the classification parity condition no longer holds in the classifier performance.

From Bayesian decision theory perspective, one fundamental reason of the bias predictions in ML models comes from the intrinsic disparities of the score distributions from the given classifier among all the groups. To achieve fairness in ML, one feasible solution is to alleviate the discrepancy of the score distributions in different groups. Once the distributions are equalized, the classifier performances are identical and indistinguishable among all the groups, whatever the decision threshold is. We propose the definition of equalized distribution as follows.

Definition 2 (Equalized Distribution) A binary classifier C achieves equalized distribution, if the score distributions are identical among all the groups $D_k, k = 1, 2, \ldots, K$, i.e.,

$$\text{pdf}(C(D_i)) = \text{pdf}(C(D_j)), \ \forall i, j = 1, 2, \ldots, K \tag{5}$$

where pdf stands for probability density function, which computes the data distribution.

More strictly, equalized distribution enforces the equalization of the group-wise score distributions from the classifier. One implication is that equalized distribution makes the prior art of the classification parity, such as demographic parity, equalized odds, always holds true whatever the decision threshold is. Hence, unlike the prior art of the parity notions, "equalized distribution" is *threshold-invariant*. The equalized distribution requires the classifier to offer equivalent and indistinguishable learning performance statistics among all the groups $D_k, k = 1, 2, \ldots, K$ independent of the decision threshold. To contrast the notion of the prior art of the classification parity and the proposed equalized distribution, we refer to the prior art as the weak condition of classification parity, and equalized distribution as the strong condition.

Based on the classification parity, we propose the fairness ML algorithms in two possible scenarios: with and without prior knowledge of the protected attribute in test stage.

4.2 With Prior Knowledge of the Protected Attribute

In this scenario, we assume that we have the prior knowledge of the protected attribute of the input samples in the test stage. We enforce equalized odds (i.e., weak classification parity) on the classifier. Given a trained classifier C, the condition of equalized odds can be satisfied by tuning the decision thresholds for different groups. We denote the number of correct predictions in group $D_k, k = 1, 2, \ldots, K$ with the decision threshold t_k as $N(D_k, t_k)$. Then the classification accuracy can be written as

$$E_a = \frac{\sum_{k=1}^{K} N(D_k, t_k)}{\sum_{k=1}^{K} |D_k|} \tag{6}$$

To achieve the equalized odds, we tune the thresholds $t_k, k = 1, 2, \ldots, K$ to shrink the discrepancy of TPR and FPR among all the groups, which can be represented as the fairness term

$$E_f = \sum_{k=2}^{K} \left(|\text{TPR}(D_1, t_1) - \text{TPR}(D_k, t_k)| + |\text{FPR}(D_1, t_1) - \text{FPR}(D_k, t_k)| \right) \tag{7}$$

where $\text{TPR}(D_k, t_k)$ and $\text{FPR}(D_k, t_k)$ denote true positive rate and false positive rate in group D_k with the decision threshold t_k. In practice, we trade-off the classification accuracy and the fairness term. Hence, given the trained classifier C, the fairness problem becomes an optimization problem

$$\max_{t_k, k=1, 2, \ldots, K} E_a - \lambda E_f \tag{8}$$

where λ is a hyperparameter. Since Eq. (8) is a nonlinear and non-differentiable problem, we apply particle swarm optimization (PSO) [14] to find the best set of thresholds $\{t_k\}_{k=1,2,\ldots,K}$ for K groups in terms of reducing the difference of TPR and FPR between two groups.

4.3 Without Prior Knowledge of the Protected Attribute

In this scenario, we assume that no prior knowledge of the protected attribute of the input samples is provided in the test stage. Different from Sect. 4.2, we enforce strong classification parity on the trained classifier. The equalized distribution condition sufficiently guarantees that the classifier can achieve classification parity among all the groups using one universal and non-specific threshold.

How can we formulate the constraint in terms of equalized distribution? We propose to approximate the continuous score distributions using the discrete histograms. Denote the sample x_i and the corresponding decision score s_i provided by the given classifier C, i.e., $s_i = C(x_i)$. The count of the scores in the bin $(c - \frac{\Delta}{2}, c + \frac{\Delta}{2})$ can be expressed as

$$n_c = \sum_i rect_c(s_i) \tag{9}$$

where c, Δ are the center and the bandwidth of the bin, and $rect_c(\cdot)$ is a rectangular function

$$rect_c(s) = \begin{cases} 1, & s \in (c - \frac{\Delta}{2}, c + \frac{\Delta}{2}) \\ 0, & o.w \end{cases} \tag{10}$$

Since such function is not differentiable, we approximate it with Gaussion function, i.e.,

$$n_c = \sum_i gauss_c(s_i) = \sum_i \exp\left(-\frac{(s_i - c)^2}{2\sigma^2}\right) \tag{11}$$

Thus, the histogram of the decision score in group D_k can be expressed as

$$hist(D_k) = \sum_c \sum_{x \in D_k} gauss_c(C(x)) \tag{12}$$

We normalize the histogram as

$$norm\_hist(D_k) = \frac{1}{|D_k|} hist(D_k) \tag{13}$$

We use L-2 loss to constraint the equalization of the score distributions in two groups, i.e.

$$E_f = \sum_{k=2}^{K} \Big(||norm\_hist(D_{1,p}) - norm\_hist(D_{k,p})||_2^2$$

$$+ ||norm\_hist(D_{1,n}) - norm\_hist(D_{k,n})||_2^2 \Big) \tag{14}$$

where $D_{k,p}$ denotes the set of positive samples in group D_k and $D_{k,n}$ denotes the set of negative samples in D_k. Like Sect. 4.2, we formulate the minimization problem, trade-off the classification accuracy and the fairness term.

$$L = \alpha E_a + (1 - \alpha) E_f \tag{15}$$

where E_a is the accuracy term, such as mean square error (MSE) or logistic loss, and $\alpha \in [0, 1]$. As the loss function L is continuous, gradient based methods can be employed to search the solution.

4.4 Experimental Results

We conducted the experiments on the COMPAS dataset [7]. The COMPAS records the recidivation of 7214 criminals in total. Each line of record documents the criminal's information of race, sex, age, number of juvenile felony criminal charges (juv_fel_count), number of juvenile misdemeanor criminal charges (juv_misd_count), number of non-juvenile criminal charges (priors_count), the previous criminal charge (charge_id), the degree of the charge (charge_degree) and the ground truth of 2-year recidivation (two_year_recid). In the experiments, we chose race as the protected attribute and, for simplicity, only focused on two race groups, "white" and "black", excluding the records of other races. Hence, we totally used 6150 samples to demonstrate the effectiveness of the proposed fairness ML algorithms in two groups. In the experiments, we randomly splited the dataset into 60% training, 20% validation, and 20% test.

4.4.1 Feature Preprocessing

Among the unprotected features of one criminal record, binary features, i.e., sex and charge_degree, are employed 0–1 encoding. Continuous features, i.e., age, juv_fel_count, juv_misd_count and priors_count are applied with data standardization. The categorical feature, charge_id, is encoded with one-hot encoding, leading to a 430-D feature vector. Thus, the total dimension of the feature vector is 436.

Then, we applied principle component analysis (PCA) to reduce the 436-D feature vector to 20-D feature vector.

According to the protected feature (e.g. race), the dataset D is divided into the "white" group D_w and the "black" group D_b. Our goal is to obtain a binary classifier to offer fair predictions of the 2-year recidivation in D_w and D_b.

4.4.2 With Prior Knowledge

We investigated two machine learning models in this scenario: logistic regression and SVM. We first trained the model with the training set as normal, and then tuned the decision thresholds on the validation set. The finalized classifier was tested on the test set. During the tuning, we set $\lambda = 1$ in Eq. (8) to trade-off the accuracy and the classification parity.

Table 5 presents the performance of the classifiers before and after tuning. Before tuning, the logistic regression classifier has the default decision threshold 0.5, while the SVM classifier has the default decision threshold 0. According to the statistics of TPR and FPR before tuning, we can see both of the classifiers are positively biased in the group "black" to "white". The tuning algorithm is employed on the validation set to shrink the discrepancy between two groups. Figure 13 shows the state of the classifier on the ROC curves before and after tuning on the validation set. We can see that the tuning algorithm help two points merge to an overlap point on the curves to fulfill the classification parity. After the tuning algorithm on the validation set, the gaps of TPR and FPR on test set between two groups were successfully reduced from around 0.2 to no more than 0.05, but the drop of the overall accuracy is insignificant. Specifically, the logistic regression classifier only decreases 1.7% on classification accuracy and the SVM classifier lowers the accuracy by 2.3%.

To sum up, the proposed tuning method enforces the fairness by tuning the decision thresholds of one given model for all the groups. It can be applied after any trained classifiers without modified them. On the other hand, the main disadvantage is that it requires access to the protected attribute of the samples in test stage. Although the tuning method can, to some extent, alleviates the biased prediction in ML model, the model itself still has distinguishable learning performances in two groups in terms of the score distributions.

Table 5 The performance of the classifiers before and after tuning

Tuning		Logistic				SVM			
		THR	Acc	TPR	FPR	THR	Acc	TPR	FPR
Before	White	0.5	67.9%	0.407	0.178	0	67.6%	0.346	0.143
	Black	0.5		0.716	0.337	0		0.620	0.252
After	White	0.503	66.2%	0.401	0.160	−0.281	65.3%	0.456	0.205
	Black	0.588		0.445	0.147	0.300		0.478	0.198

THR decision threshold, *Acc* total classification accuracy, *TPR* true positive rate, *FPR* false positive rate

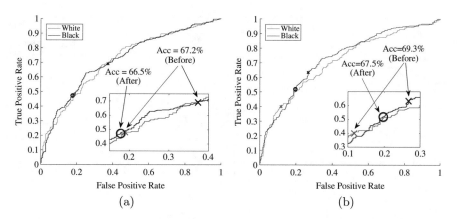

Fig. 13 The state of the classifier on the ROC curves on the validation set. The Crosses represent the states before tuning, and circles represent the states after tuning. (**a**) Logistic regression classifier. (**b**) SVM classifier

4.4.3 Without Prior Knowledge

In this scenario, we conducted experiments with logistic regression classifier. The accuracy term E_a in Eq. (15) is the logistic loss for binary classification, i.e.,

$$E_a = \sum_i y_i \log C(x_i) + (1 - y_i) \log(1 - C(x_i)) \tag{16}$$

where C is the trained classifier, and (x_i, y_i) is the sample and label pair in the whole dataset D. We set the bandwidth of the bin in the histogram Δ as 0.02. Since the output of the logistic regression classifier is confined in the range [0, 1], the center of the bins are 0.01, 0, 03, 0.05, ..., 0.97 and 0.99. In the training, we applied momentum gradient descent to optimize the loss function in Eq. (15), with learning rate $\mu = 1$ and momentum $= 0.9$. The number of learning iteration is 2k.

We conducted the experiments with different values of the hyperparameter α to explore the influence of α on the performance of the classifier, which is shown in Fig. 14. When α is 1, the classifier is the normal classifier with minimum error. When α is close to 0, the classifier is a mapping to equalize the score distributions in two groups, emphasizing the strong classification parity. When α increases from 0 to 1, weighing more on the accuracy loss, the accuracy of the classifier first increases significantly and then saturates to an upper bound after $\alpha = 0.2$. As for the classification parity, the differences of TPR/FPR between two groups also enlarge as α increases, indicating the deteriorate the fairness in ML classifier. Besides, the score distributions of the white and the black in test set with different α values are presented in Fig. 15. We can see that the smaller α, the closer the score distributions of the positive/negative samples in "white" and "black" groups. Balancing the accuracy and fairness, we recommend that the best range of α is 0.1–0.2.

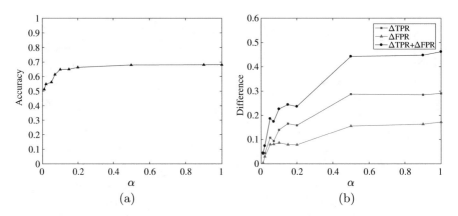

Fig. 14 The influence of the hyperparameter α on the performance of the classifier. (**a**) The overall accuracy of the classifier with different α. (**b**) The fairness indicator of the classifier with different α, i.e., the difference of TPR/FPR between two groups

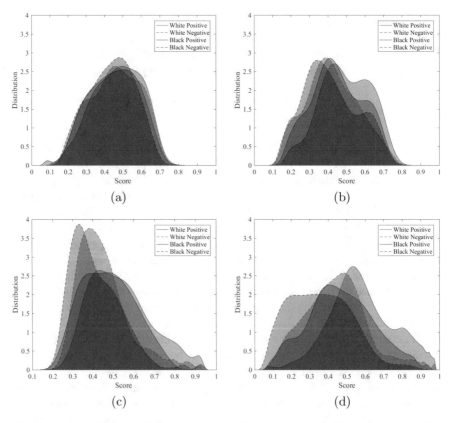

Fig. 15 The score distributions of the white and the black in test set with different α, (**a**) $\alpha = 0.01$, (**b**) $\alpha = 0.1$, (**c**) $\alpha = 0.2$, and (**d**) $\alpha = 1$

From the experiments, we can see that the proposed training method can provide a classifier equalizing the distribution of the output scores among the groups. Ideally, the classifier has equal and indistinguishable learning performance statistics, e.g., TPR and FPR, among all the protected attributes.

Acknowledgments This work was initiated at the Workshop for Women in Math and Public Policy. We would like to thank IPAM and The Luskin Center for hosting the workshop as well as the organizers Mary Lee and Aisha Najera Chesler for their tireless efforts. We also thank the anonymous reviewers for their constructive comments, which helped to significantly improve the presentation.

References

1. Solon Barocas, Moritz Hardt, and Arvind Narayanan. *Fairness and Machine Learning*. fairmlbook.org, 2019. http://www.fairmlbook.org.
2. Solon Barocas and Andrew D. Selbst. Big Data's Disparate Impact. *SSRN eLibrary*, 2014.
3. Toon Calders, Faisal Kamiran, and Mykola Pechenizkiy. Building classifiers with independency constraints. In *ICDM Workshops 2009, IEEE International Conference on Data Mining Workshops, Miami, Florida, USA, 6 December 2009*, pages 13–18, 2009.
4. Flavio Calmon, Dennis Wei, Bhanukiran Vinzamuri, Karthikeyan Natesan Ramamurthy, and Kush R Varshney. Optimized pre-processing for discrimination prevention. In *Advances in Neural Information Processing Systems*, pages 3992–4001, 2017.
5. Sam Corbett-Davies and Sharad Goel. The measure and mismeasure of fairness: A critical review of fair machine learning. *arXiv preprint arXiv:1808.00023*, 2018.
6. New York Police Department. Stop, Question and Frisk Data. https://www1.nyc.gov/site/nypd/stats/reports-analysis/stopfrisk.page.
7. Julia Dressel and Hany Farid. The accuracy, fairness, and limits of predicting recidivism. *Science Advances*, 4(1):eaao5580, 2018.
8. Cynthia Dwork, Moritz Hardt, Toniann Pitassi, Omer Reingold, and Richard S. Zemel. Fairness through awareness. In *Innovations in Theoretical Computer Science 2012, Cambridge, MA, USA, January 8–10, 2012*, pages 214–226, 2012.
9. Cynthia Dwork, Nicole Immorlica, Adam Tauman Kalai, and Max Leiserson. Decoupled classifiers for group-fair and efficient machine learning. In *Conference on Fairness, Accountability and Transparency*, pages 119–133, 2018.
10. Michael Feldman, Sorelle A. Friedler, John Moeller, Carlos Scheidegger, and Suresh Venkatasubramanian. Certifying and removing disparate impact. In *Proceedings of the 21th ACM SIGKDD International Conference on Knowledge Discovery and Data Mining, Sydney, NSW, Australia, August 10–13, 2015*, pages 259–268, 2015.
11. James Foulds and Shimei Pan. An intersectional definition of fairness. *arXiv preprint arXiv:1807.08362*, 2018.
12. Moritz Hardt, Eric Price, and Nati Srebro. Equality of opportunity in supervised learning. In *Advances in neural information processing systems*, pages 3315–3323, 2016.
13. Matthew Joseph, Michael J. Kearns, Jamie H. Morgenstern, and Aaron Roth. Fairness in learning: Classic and contextual bandits. In *Advances in Neural Information Processing Systems 29: Annual Conference on Neural Information Processing Systems 2016, December 5–10, 2016, Barcelona, Spain*, pages 325–333, 2016.
14. James Kennedy. Particle swarm optimization. *Encyclopedia of Machine Learning*, pages 760–766, 2010.
15. Jon Kleinberg, Sendhil Mullainathan, and Manish Raghavan. Inherent trade-offs in the fair determination of risk scores. *arXiv preprint arXiv:1609.05807*, 2016.

16. Ya'acov Ritov, Yuekai Sun, and Ruofei Zhao. On conditional parity as a notion of non-discrimination in machine learning. *arXiv preprint arXiv:1706.08519*, 2017.
17. Sakshi Udeshi, Pryanshu Arora, and Sudipta Chattopadhyay. Automated directed fairness testing. In *Proceedings of the 33rd ACM/IEEE International Conference on Automated Software Engineering, ASE 2018, Montpellier, France, September 3–7, 2018*, pages 98–108, 2018.
18. Hao Wang, Berk Ustun, and Flávio P. Calmon. Repairing without retraining: Avoiding disparate impact with counterfactual distributions. In *Proceedings of the 36th International Conference on Machine Learning, ICML 2019, 9–15 June 2019, Long Beach, California, USA*, pages 6618–6627, 2019.
19. Muhammad Bilal Zafar, Isabel Valera, Manuel Gomez Rodriguez, and Krishna P Gummadi. Fairness beyond disparate treatment & disparate impact: Learning classification without disparate mistreatment. In *Proceedings of the 26th International Conference on World Wide Web*, pages 1171–1180. International World Wide Web Conferences Steering Committee, 2017.
20. Muhammad Bilal Zafar, Isabel Valera, Manuel Gomez Rodriguez, and Krishna P Gummadi. Fairness constraints: Mechanisms for fair classification. In *20th International Conference on Artificial Intelligence and Statistics*, pages 962–970, 2017.
21. Rich Zemel, Yu Wu, Kevin Swersky, Toni Pitassi, and Cynthia Dwork. Learning fair representations. In *International Conference on Machine Learning*, pages 325–333, 2013.
22. Richard S. Zemel, Yu Wu, Kevin Swersky, Toniann Pitassi, and Cynthia Dwork. Learning fair representations. In *Proceedings of the 30th International Conference on Machine Learning, ICML 2013, Atlanta, GA, USA, 16–21 June 2013*, pages 325–333, 2013.
23. Indre Zliobaite. On the relation between accuracy and fairness in binary classification. *CoRR*, abs/1505.05723, 2015.

The Hidden Price of Convenience: A Cyber-Inclusive Cost-Benefit Analysis of Smart Cities

Ummugul Bulut, Emily Frye, Jillian Greene, Yuanchun Li, and Mary Lee

Abstract The vision and technologies for Smart Cities (also called Connected Cities) promise to improve public services and quality of life by using different types of software, hardware, networking, and analysis to collect and analyze data. Many practical and economic benefits appear likely when officials explore Smart Cities and their enabling technologies. However, research shows that, to date, Smart Cities are not requiring cybersecurity during design, acquisition, testing, development, deployment, and operation. The result is that Smart Cities are actually incorporating additional risk to essential functions that cities perform for citizens. The authors demonstrate that, at this time, Smart Cities are not accurately accounting for risk in the assessments that they make prior to moving in the direction of Smart City status. In this paper, a reasonably foreseeable, hypothetical scenario for a medium-sized city is described and analyzed in order to show why more complete cost-benefit analysis is required for Smart Cities and officials pursuing Smart City capability.

U. Bulut (✉)
Texas A&M University, San Antonio, TX, USA
e-mail: ubulut@tamusa.edu

E. Frye
Mitre Corporation, McLean, VA, USA
e-mail: fefrye@mitre.org

J. Greene
Washington University in St. Louis, St. Louis, MO, USA
e-mail: JillianGreene@wustl.edu

Y. Li
University of Southern California, Los Angeles, CA, USA
e-mail: yli022@usc.edu

M. Lee
RAND Corporation, Santa Monica, CA, USA
e-mail: mlee@rand.org

© The Author(s) and the Association for Women in Mathematics 2020   81
M. Lee, A. Najera Chesler (eds.), *Research in Mathematics and Public Policy*, Association for Women in Mathematics Series 23, https://doi.org/10.1007/978-3-030-58748-2_5

1 Background and Introduction

Cities need to operate as efficiently as possible. Generally, they are on the alert when cost-saving innovations appear in the marketplace.

It is not surprising that the apparent operational efficiencies and cost savings of increased automation appeal to urban managers. Conceptually, cities can reap the benefit from becoming "smart": moving toward equipment and infrastructure that enables remote management and monitoring by a few employees saves on personnel costs and provides many years of depreciable functionality. While there is no canonical definition of Smart City (interchangeably called a Connected City), certain elements form a common basis. Smart Cities use connected devices, often sensors, to collect and aggregate data pertaining to essential city functions and citizen services (such as power, traffic, or first-responder deployment). These data are analyzed and used to inform decisions about operating, adjusting, or increasing resources related to these functions and services. Often, some of the sets of data are consolidated in one or more central review and decision offices. They may be called Emergency Operation Centers or Security Operations Centers.

At the same time, the authors assert that Life-cycle Cost Estimates for Smart Cities today are fatally flawed. Why? Because they fail to incorporate the cybersecurity design, configuration, maintenance, and upgrade costs that are essential for basic protection of the kinds of functions that Smart City technologies enable; and they fail to account for the costs of the consequences when poorly secured systems are, inevitably, compromised. The literature has not explored this issue.[1]

Today, Smart Cities focus on the benefits of automating a range of functions, including traffic management, water usage, benefits distribution, and emergency dispatch. Almost all of these functions are critical to public health and safety. Therefore, if something undermines these functions, public health and safety are at risk.

Given the stakes associated with Smart City functionality, a thorough evaluation of benefits and costs is essential for cities seeking to make rational purchase and policy decisions. Do city managers, mayors, and others today have adequate insight into the risks and costs of deploying Smart City capability? If the true costs of a cyberattack on Smart-City functions are understood, do city managers make different decisions? Do vendors change their cybersecurity posture for embedded systems? Do acquisition professionals and grantmakers specify requirements so that cybersecurity risks are explicitly addressed?

The authors surveyed existing literature and market developments, as well as grant requirements for Smart City development in a small set of identified grants.

---

[1]One of the primary grant recipients from the National Science Foundation is dedicated to the development and deployment of new and greater smart city capability across the nation, but consideration of cybersecurity and/or an assessment of cyber-related risk is not included in the program.

Universally, cybersecurity was not addressed or was addressed in general, vague terms.

> The threat of cybersecurity compromise is neither theoretical nor vague. as the authors conclude edits, the city of New Orleans has declared a state of emergency caused by cyberattack. In the past year alone, city government function has been compromised in Atlanta, multiple counties in Massachusetts have effectively been crippled by ransomware, and much of Baltimore's government functionality was stalled for weeks. Through August of 2019, twenty-three jurisdictions in Texas are crippled by ransomware—in localities that have not intentionally pursued Smart City functionality. Clearly, we are not thinking rationally about the costs of Smart Gone Wrong.

These are not incidents that cities welcome. Now, before cities make additional purchase and deployment decisions about long-lived systems for essential public safety functions, is the time to assess the risk and actual cost of poor cybersecurity.

The January 2019 Women in Math and Public Policy Workshop allowed the Cyber Team an opportunity to focus on this fundamental problem in the evolution toward Smart Cities. The team analyzed reasonably foreseeable outcomes associated with cybersecurity compromise to functions that are deployed in cities now or that are in the process of deployment. Cost-benefit analysis for Smart City functions has not, to the best of the team's extensive knowledge and research, incorporated cybersecurity requirements, costs, or risks. The team explored and documented initial steps toward establishing an understanding of why Cost-Benefit Analysis (CBA) must include costs of cybersecurity and risks associated with cyber compromise.

This article summarizes the team's work. The following sections describe:

- A reality-based hypothetical scenario that the team developed so as to understand how invisible cybersecurity risks are likely to appear in Smart Cities
- The scenario timeline
- Discussion of costs versus benefits
- Research process and resources

The team does not represent that this is an exhaustive data collection or assessment of cybersecurity risks and approaches in Smart Cities. Instead, the study is illustrative; the authors' aim is to open up the national dialogue about an area that has been opaque and to provide an exploratory foundation upon which we hope others will build.

2 A Hypothetical Catastrophe

As a result of the aforementioned threat landscape, our research group decided to explore one medium-sized mid-country city and to apply a reasonably foreseeable disaster scenario against the city's current and planned capabilities. Our hypothesized city, which we will call Central City, mimics the geography and demography of an actual city that has received a large federal grant to develop Smart-City capabilities. Assumptions in the scenario are based on this actual urban area.

The geographic clustering within Central City is not an unusual layout for a city in the United States [2]; daytime high population toward the central downtown area is common in urban environments. In Fig. 1, the population density distribution of Central City as of 2018 is shown with the increasing darkness representing the increasing density of the area. The particularly dense patch in the center of the city is where the Central City University (CCU) campus is located.

Because of this concentrated congestion, a large part of the aim of Central City's Smart City vision is to create a highly connected traffic flow system that provides real-time data and that allows light timing and patterns to be altered remotely. The city has several plans in this regard, including the following excerpt from their grant application [3]:

Fig. 1 Population density of Central City [2]

Central City has completed construction of a new state-of-the-art Traffic Management Center (TMC), and construction is presently underway with the city's multi-year, $76 million investment in the Central Traffic Signal System (CTSS) project. When completed in 2018, CTSS will link the Central City TMC to all 1,250 signalized Central City intersections and utilize new weather-tracking software and sensors to relay pavement conditions while a GPS system provides real-time information about where snowplows are operating. Most importantly, the CTSS project and the new Central City TMC will provide traffic management coordination between Central City, 12 regional communities, Central City County, CCU and the state-wide department of transportation. The TMC consolidates traffic signal, special event traffic, and snow removal operation command into a centralized location for the city. Central City is also home to the department of transportation's TMC, which monitors traffic conditions in each of the major metropolitan areas of the state and is linked to city infrastructure through a robust network of sensors, cameras and communication technologies.

Reducing traffic congestion and delays is a crucial issue for cities to tackle; not only is it inconvenient to the drivers, but these delays are actually having a measurable impact on the economies of several U.S. metropolitan regions [15]. While this budding emergence of connected cities is believed to be the answer to the overwhelmingly frequent and often unnecessary delays caused by this traffic congestion, there remains a very clear risk underlying the ability to control such vital utilities remotely. The application with which Central City won the competition very thoroughly details how it will protect each citizen's personal data, but nowhere does the application [3] mention cybersecurity, or a potential lack thereof. Upon looking through the applications from the 72 real-world applicants, it was evident and verifiable that the cybersecurity threats were either not obvious or not important to the proposal writers, as very few submissions even mentioned network security. To combat this blatant disregard or fundamental lack of awareness, our research group concluded that a first step is simply illustrating a potential scenario where the connectedness of a city could (a) increase the effectiveness of a terrorist attack and (b) increase rather than decrease the costs of citizen services in a Smart City.

2.1 A Concocted Catastrophe

Connected cities are a new concept. Consequently, no one can cite examples of effects—positive or negative—that follow from instituting these measures. Accordingly, all that policymakers, engineers, and computer scientists can do is try to brainstorm some realistic, educated-but-speculative disasters and proceed to anticipate reasonable risk mitigation measures. This notion is precisely what guided our group to invent a scenario for a combined cyber/physical terrorist attack on the terrain known as Central City. In this scenario, the Smart City features are all successfully installed and working for the city. Particularly, the street lights can be controlled remotely. The terrorists have infiltrated the network and now have access to all of the traffic signals in and around the city. An assumption is that it is rush hour on a Tuesday morning when university classes are in session.

First, the terrorists plant three bombs in the most densely populated regions of Central City. In doing our research for this part of our analysis, we considered the infamous Oklahoma City bombings [6], as they were made with relatively easy to find materials (fertilizer, diesel fuel, etc.), were detonated on US soil, and had a measurable impact. Next, the attackers make several fake calls about a live bomb to 911 and to the Emergency Operations Center, explicitly threatening harm to people at Central City International Airport (CCIA), located inside the red box in Fig. 2. After a significant number of the city's emergency first responders attend to these very serious claims they are receiving, the attackers shut down the traffic signals across the city. This quickly causes congestion throughout the area's primary arteries. Secondary arteries absorb overflow and also become backed up. At this time, the adversaries detonate the three bombs. The bombs are represented by red circles in and around the University– one of the farthest urban locations in relation to the airport, additionally made difficult to access because of traffic congestion. The hypothetical attack was planned for a busy weekday morning, when many commuters are on their routes to work, as we can see in the Table 1:

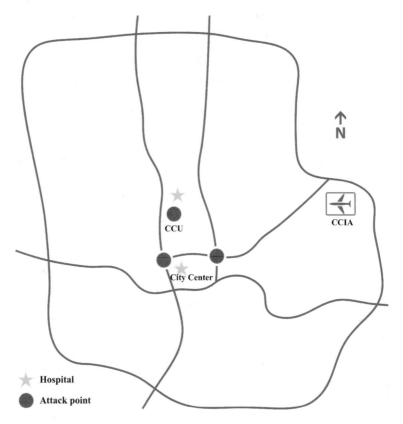

Fig. 2 Central City map and key landmarks in the terrorist attack

Table 1 Hypothetical cyber/physical terrorist attack timeline

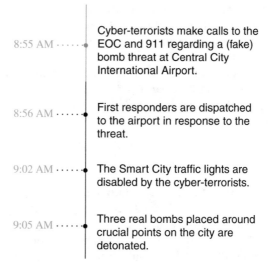

8:55 AM ·····• Cyber-terrorists make calls to the EOC and 911 regarding a (fake) bomb threat at Central City International Airport.

8:56 AM ·····• First responders are dispatched to the airport in response to the threat.

9:02 AM ·····• The Smart City traffic lights are disabled by the cyber-terrorists.

9:05 AM ·····• Three real bombs placed around crucial points on the city are detonated.

The careful orchestration involved in this plan is akin to other such plans, with the simple addition of compromise to cyber-enabled Smart City functions. The combined cyber and physical attack vectors allow a relatively unsophisticated attacker to cause significant damage—in lives, injuries, and public confidence.

Later in the paper, we discuss the suspected losses more analytically—in lives, capital, and trust—and address further risks. Although our numbers are merely extrapolated and compounded from various preexisting data sources (rather than collected as a result of empirical research for this specific study), the goal is to demonstrate the nature of the problem.

3 Analysis and Mathematical Approach

There are various economic costs associated with the types of incidents described in the instant scenario. These costs can be classified as direct and indirect costs. Direct costs might include the value of the damaged or destroyed equipment, housing, or structure as well as lost lives, injured people and associated medical expenses. Indirect costs, most difficult to measure, might include the pain and suffering of the victims and their relatives as well as negative economic impact on the travel and tourism industry [4]. In this section, the cost function is generated based on the injured individuals' health expenses and the cost of the lost lives. However, readers should realize if other components of the direct and indirect costs are included in the calculation, the cost would be higher; terrorist attacks are intended to leverage violence to create a psychological impact far outweighing the physical damage from the attack. These aspects of cost are harder to measure, and the team did not want to argue the point in an unsettled area of cost assessment [5]. In order to bound the scope of the analysis under the time constraints of the workshop, the team focused on a specific economic question: the comparative evaluation of medical costs due to treatment delays incurred by a cyber attack in smart and non-smart cities, as described in Sect. 2. The cost function, given by Eq. (1), characterizes the cumulative dollar costs, C, of medical expenses incurred by delayed response times and traffic congestion.

$$C(x, \alpha, \beta, k) = \sum_{i=1}^{n} (x - \beta_i k_i) p \alpha_i \tag{1}$$

The cost C is a function of: the total number of people, x, injured during an incident; the total number of ambulance drop-offs, n, during the incident; the waiting time α for each consecutive drop-off; the fixed cost p for each minute delay in treatment for each injured person; the total number of people, β, in each ambulance at drop-off; and the total number of ambulances, k, at each location. Total number of drop-offs, n, can be calculated by using the following formula:

$$n = \frac{x}{\beta k} \tag{2}$$

Based on Ref. [14], the national average emergency response time for big/medium city is 8 min. Given deployment of first responders to the other side of the city from the bomb locations, the non-smart city response is assumed to double. In a smart city, response time is assumed longer by 3 min than response times in a regular city: it gets even higher than the regular city since attackers/hackers are able to remotely shut down, as a result of the intentional connectivity of a smart city, the network of sensors such as cameras, wireless devices, traffic lights, etc. Traffic congestion climaxes 15–25 min post-detonation. Traffic congestion will result in long delays for EMTs, so the medical cost will distinctly increase since 1 min delay in the treatment of an injured patient costs $5000 in medical bills [11]. There are different emergency response mobilization methods discussed in the literature [1]. For the current catastrophe, multiple emergency responses and evacuation flow groups with different destinations are assumed because of the assumed three bombs planted in several different areas of Central City.

For the above scenario, 140 people are assumed to be injured at three different locations. Incident-related costs, specifically hospital expenses, are calculated for smart and non-smart cities separately. For both of these cities at 8:55 a.m., the EOC and 911 receive fraudulent calls about the bombing at the airport which is located on the opposite side of the city from locations where attackers plan to detonate the real bombs. At 9:05 a.m. terrorists explode real bombs in three different locations simultaneously. In a non-smart and smart city, EMTs arrive at three incident locations at 9:21 and 9:24 a.m. respectively. In a smart city, the emergency response time increases as attackers turn on and off traffic lights, to increase traffic congestion. In a non-smart city, the first dispatch of EMTs consists of seven ambulances and arrives at the closest hospital at 9:31 a.m. Ambulances can carry up to three people depending on the severity of their injuries, thus there would be 119 injured people left at three different incident locations. The next dispatch occurs at 9:35 a.m., so as of 9:31 a.m. the cost associated with the treatment delay of 119 people can be calculated as $2,380,000. On the other hand, for the smart-city, the first dispatch of EMT occurs at 9:34 a.m. and the next dispatch occurs at 9:39 a.m. The number of ambulances and their carrying capacities are assumed to be same for both of these cities. Thus in a smart-city, as of 9:34 a.m., 5 min delay costs $2,975,000. The treatment-delay cost between smart and non-smart cities accumulates and eventually the cost difference increases dramatically; for the particular scenario in a smart city delay in the treatment costs $4,280,000 more than the non-smart city. The time series graphs in Fig. 3 show the accumulated cost at each dispatch time for smart and non-smart cities. The cost difference increases as the timeline continues, due to more fatalities or injured people. Total health related costs in non-smart and smart cities are calculated as $8,720,000 and $13,000,000 respectively. Notice that for both of these graphs, the initial time is set to be 9:31 a.m. which is the first dispatch of EMT in non-smart city. This time is considered to be the reference point for smart city as well for comparison purposes.

4 Research Process and Resources

The embryo of the Smart City pattern is generated from the background that a specific city won the competitive grant application Smart City Challenge against 77 cities in 2016 [7]. We focused on this city as our research subject, on account of its Smart City Operating System (SCOS) which can support measuring progress and performance of U.S. Department of Transportation's (USDOT) grant initiatives [7]. SCOS is a web-based data platform which serves to collect and distribute single points of data [8]. A connected vehicle environment installs connection between cars and with city infrastructure. However the density of data-based applications

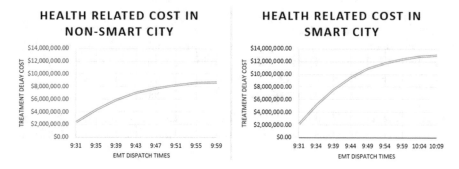

Fig. 3 Cost function graph for smart and non-smart city

in the city transportation increased the opportunity surface for cyber adversaries. The hypothesis used here draws as well on the Dangers of Smart City Hacking whitepaper [9], Distributed denial of service (DDoS) anonymous attack protocols and massive reports of cyber attack paralyzing the city utility system [13].

Initially, surrounding the central data layer, a toy model of the city infrastructure is built in Gephi, an open-source visualization software. Primary nodes of the sensor, traffic light, data hub are directionally bonded. As the next step, the realistic scenario-based analysis of an attack is deduced.

Under the scenario used here, the Central Smart City operating system is the object of a cyber-attack which infiltrates into the whole traffic light network; a physical attack includes three forceful bombs placed in three densely populated landmarks. The timeline of attack and emergency action was assumed. The emergency response time for a small/medium city is assumed to be 8 min for EMT and firefighters, based on the following performance indicators report [10]. the effect of late response time on mortality is also a consideration [11].

Cost-benefit analysis (CBA) provides a basis by which to determine if the good of project in social benefits exceeds the costs associated with it [12]. Thus we estimated the cost on account of delayed rescue as above. According to the preliminary estimates, the emergency response waiting times alone will add more than 4 million dollars to city burden.

Beyond the cost of waiting time, additional costs would be incurred. These costs include long term medical treatment/physical rehabilitation costs for injuries, additional traffic collisions after traffic light outage, software and hardware reconstruction costs and so on. In our further research, these costs will be calculated and included in CBA of smart cities versus not smart cities as well.

5 Conclusion

Smart Cities offer the lure of potential enhanced speed and efficiency. At the same time, they create different risks from those that cities historically have faced.

The analysis in this article is presented to illustrate the kinds of considerations that help urban planners, city leaders, and first responder communities understand these different risks. City leaders need to define and explore their risk tolerance and resilience options before investing in Smart City tools and capabilities. In order to clearly and fully assess lifecycle costs and critical-functions risks that come with the purchase and installation of new technologies, we recommend a pause for cyber risk assessment. Cities can consider the following approaches:

- A community of intentional vulnerability disclosure and remediation should be established that covers Smart City software, hardware, and common configuration.

- DHS and/or NIST should develop reference architectures, best practices guidance, acquisition language, sample cybersecurity requirements, and obsolescence and decommissioning guides.
- Vendors must provide patchable systems.
- The Smart City vendor community must be held responsible for educating city managers/operating personnel in proper configuration and operation for cybersecurity protocols, not simply system function.

6 Future Research

A range of cities needs to be studied to evaluate best practices in designing, procuring/acquiring (including validation during source selection), installing, and monitoring Smart City capabilities. DHS is a logical leader for future analysis and best practice development.

Acknowledgments Authors are grateful to the organizers of 2019 Women in Math and Public Policy Workshop as well as IPAM for hosting group members at UCLA campus.

References

1. Chiu, J., Zheng H.: Real-time mobilization decisions for multi-priority emergency response resources and evacuation groups: Model formulation and solution. TRANSPORT RES E-LOG. (2007). https://doi.org/10.1016/j.tre.2006.11.006
2. Dod, J.: in *Where Are They Going? Population Growth in Franklin County, Ohio* (2013) Available via DIALOG. http://urbandecisiongroup.com/where-are-they-going-population-growth-in-franklin-county-ohio/respond of subordinate document. Cited 1 Jan 2019
3. Dod, J.: in *Ohio Department of Transportation Columbus Smart City Application* (2016) Available via DIALOG. https://www.transportation.gov/sites/dot.gov/files/docs/Columbus%20OH%20Vision%20Narrative.pdf of subordinate document. Cited 6 Jun 2019
4. Dod, J.: in *Measuring the Economic Costs of Terrorism* Columbus Smart City Application (2016) Available via DIALOG. https://www.socsci.uci.edu/~mrgarfin/OUP/papers/Enders.pdf of subordinate document. Cited 6 Jun 2019
5. Dod, J.: in *Understanding and Responding to Bomb Threats* (2016) Available via DIALOG. https://worldview.stratfor.com/article/understanding-and-responding-bomb-threats of subordinate document. Cited 6 Jun 2019
6. Dod, J.: in *Oklahoma City Bombing* (2016) Available via DIALOG. https://www.fbi.gov/history/famous-cases/oklahoma-city-bombing. Cited 30 Aug 2019
7. Dod, J.: in *About smart Columbus* (2019) Available via DIALOG. https://www.smartcolumbusos.com/about/about-smart-columbus of subordinate document. Cited 1 Jan 2019
8. Dod, J.: in *How Columbus is Accomplishing its Smart City Vision* (2018) Available via DIALOG. https://datasmart.ash.harvard.edu/news/article/how-columbus-accomplishing-its-smart-city-vision of subordinate document. Cited 1 March 2019
9. Dod, J.: in *How to Outsmart the Smart City* (2016) Available via DIALOG. http://ibm.biz/smartcities of subordinate document. Cited 1 Aug 2019

10. Dod, J.: in *2015 Operating Budget Proposal* (2015) Available via DIALOG. https://www.columbus.gov/finance/financial-management-group/budget-management/FY15-Operating-Budget-Proposal/ of subordinate document. Cited 1 Aug 2019
11. Dod, J.: in *Outcomes Quantifying the Impact of Emergency Response Times* (2015) Available via DIALOG. https://cdn2.hubspot.net/hubfs/549701/Documents/RapidSOS_Outcomes_White_Paper_-_2015_4.pdf of subordinate document. Cited 1 Aug 2019
12. Granberg, T.A., Weinholt A.: New Collaborations in Daily Emergency Response: Applying cost-benefit analysis to new first response initiatives in the Swedish fire and rescue service. Int. J. Emerg. Serv. (2015). https://doi.org/10.1108/IJES-01-2015-0002
13. Guri M., Mirsky Y., Elovici Y.: 9-1-1 DDoS: Threat, Analysis and Mitigation. (2016). https://arxiv.org/abs/1609.02353
14. Haider A.H., Haut E.R., Velmahos G.C.: Converting Bystanders to Immediate Responders: We Need to Start in High School or Before. JAMA Surg. **152**, 909–910 (2017)
15. Sweet, M.: Traffic Congestion's Economic Impacts: Evidence from US Metropolitan Regions. URST. **151**, 2088–2110 (2013)

Security of NewHope Under Partial Key Exposure

Dana Dachman-Soled, Huijing Gong, Mukul Kulkarni, and Aria Shahverdi

Abstract Recently, the work of Bolboceanu et al. (Asiacrypt '19) and the work of Dachman Soled et al. (Mathcrypt '19) have studied a leakage model that assumes leakage of some fraction of the NTT coordinates of the secret key in RLWE cryptosystems (or equivalently, intentionally sampling secrets with some fraction of NTT coordinates set to 0). This can be viewed as a partial key exposure problem, since for efficiency purposes, secret keys in RLWE cryptosystems are typically stored in their NTT representation. We extend this study by analyzing the security of the NewHope key exchange scheme under partial key exposure of $1/8$-fraction of the NTT coordinates of the parties' secrets. We adopt the formalism of the decision Leaky-RLWE (Leaky-DRLWE) assumption introduced in the work of Dachman Soled et al. (Mathcrypt '19), which posits that given leakage on a sufficiently small fraction of NTT coordinates of the secret, the remaining coordinates of the output remain indistinguishable from uniform. We note that the assumption in the work of Dachman Soled et al. (Mathcrypt '19) is strictly weaker than the corresponding assumption in the work of Bolboceanu et al. (Asiacrypt '19), which requires that the entire output remain indistinguishable from uniform. We show that, assuming that Leaky-DRLWE is hard for $1/8$-fraction of leakage, the shared key v (which is then hashed using a random oracle) is computationally indistinguishable from a random variable with average min-entropy 237, conditioned on the transcript and leakage,

Part of this work was done while the author Mukul Kulkarni was a student at the University of Maryland.

D. Dachman-Soled · H. Gong (✉) · A. Shahverdi
University of Maryland, College Park, MD, USA
e-mail: danadach@umd.edu; danadach@ece.umd.edu; gong@cs.umd.edu; ariash@umd.edu

M. Kulkarni
University of Massachusetts, Amherst, MA, USA
e-mail: mukul@cs.umass.edu

© The Author(s) and the Association for Women in Mathematics 2020　　　　93
M. Lee, A. Najera Chesler (eds.), *Research in Mathematics and Public Policy*, Association for
Women in Mathematics Series 23, https://doi.org/10.1007/978-3-030-58748-2_6

whereas without leakage the min-entropy is 256. Note that $2 \cdot 1738$ number of bits of information are leaked in this leakage model, and so the fact that any entropy remains in the shared 256-bit key is non-trivial.

1   Introduction

The cryptographic community is currently developing "post-quantum" crypto systems—cryptosystems believed to remain secure even in the presence of a quantum adversary—to replace known quantum-insecure cryptosystems based on the factoring and discrete log assumptions. One of the foremost avenues for efficient, post-quantum public key cryptography is the construction of cryptosystems from the Ring-LWE (RLWE) assumption. RLWE is often preferred in practice over standard LWE due to its algebraic structure, which allows for smaller public keys and more efficient implementations. In the RLWE setting, we typically consider rings of the form $R_q := \mathbb{Z}_q[x]/(x^n + 1)$, where n is a power of two and $q \equiv 1 \mod 2n$. The (decisional) RLWE problem is then to distinguish $(a, b = a \cdot s + e) \in R_q \times R_q$ from uniformly random pairs, where $s \in R_q$ is a random secret, the $a \in R_q$ is uniformly random and the error term $e \in R$ has small norm. A critical question is whether the additional algebraic structure of the RLWE problem renders it less secure than the standard LWE problem. Interestingly, to the best of our knowledge—for the rings used in practice and for practical parameter settings—the best attacks on RLWE are generic and can equally well be applied to standard LWE [45]. However, the situation with respect to robustness under leakage is quite different. While LWE is known to retain its security under leakage, as long as the secret has sufficiently high min-entropy conditioned on the leakage [32], the same is not always true for RLWE, as shown in several recent works [9, 21]. In this work, we explore leakage models under which RLWE-based cryptosystems *can* be proven secure.

The NTT Transform   A key technique for fast computation in the RLWE setting is usage of the *NTT transform* (similar to the Discrete Fourier Transform (DFT), but over finite fields) to allow for faster polynomial multiplication over the ring R_q. Specifically, applying the NTT transform to two polynomials $p, p' \in R_q$—resulting in two n-dimensional vectors, $\widehat{p}, \widehat{p}' \in \mathbb{Z}_q^n$—allows for *component-wise* multiplication and addition, which is highly efficient . Typically, the RLWE secret will then be stored in NTT form, and so leakage of coordinates of the NTT transform is a natural way to model partial key exposure attacks.

NewHope Key Exchange Protocol   Our results focus on analysis of the NewHope key exchange protocol of [4] in the presence of partial key exposure. Briefly, NewHope key exchange is a post-quantum key exchange protocol that has been considered as a good candidate for practical implementation, due to its computational efficiency and low communication. Specifically, Google has experimented with large-scale implementation of NewHope in their Chrome browser [14] to determine the feasibility of switching over to post-quantum key exchange in the near-term.

This Work The goal of this work is to further the study of partial key exposure in RLWE based cryptosystems, initiated in [21] and [9]. Specifically, we adopt the notion of the decisional versions of Leaky RLWE assumptions introduced in [21], where the structured leakage occurs on the coordinates of the NTT transform of the LWE secret (and/or error) and analyze the security of the NewHope key exchange protocol under the decision version of the assumption.

1.1 Leaky RLWE Assumptions–Search and Decision Versions

We next briefly introduce the search and decision versions of the Leaky RLWE assumptions.

For $p \in R_q := \mathbb{Z}_q/(x^n + 1)$ we denote $\widehat{p} := \mathsf{NTT}(p) := (p(\omega^1), p(\omega^3), \ldots, p(\omega^{2n-1}))$, where ω is a primitive $2n$-th root of unity modulo q, and is guaranteed to exist by choice of prime q, s.t. $q \equiv 1 \mod 2n$. Note that $\widehat{p}$ is indexed by the set $\mathbb{Z}_{2n}^*$.

The search version of the Ring-LWE problem with leakage, denoted SRLWE, is parameterized by $(n' \in \{1, 2, 4, 8, \ldots n\}, S \subseteq \mathbb{Z}_{2n'}^*)$. The goal is to recover the RLWE secret $s = \mathsf{NTT}^{-1}(\widehat{s})$, given samples from the distribution $D_{real,n',S}$ which outputs $(\widehat{a}, \widehat{a} \cdot \widehat{s} + \widehat{e}, [\widehat{s}_i]_{i \equiv \alpha \mod 2n' \, |\forall \alpha \in S})$, where a, s, and e are as in the standard RLWE assumption.

The decision version of the Ring-LWE problem with leakage, denoted DRLWE is parameterized by $(n' \in \{1, 2, 4, 8, \ldots n\}, S \subseteq \mathbb{Z}_{2n'}^*)$. The goal is to distinguish the distributions $D_{real,n',S}$ and $D_{sim,n',S}$, where $D_{real,n',S}$ is as above and $D_{sim,n',S}$ outputs $(\widehat{a}, \widehat{u}, [\widehat{s}_i]_{i \equiv \alpha \mod 2n' \, |\forall \alpha \in S})$, where $\widehat{u}_i = \widehat{a}_i \cdot \widehat{s}_i + \widehat{e}_i$ for $i \equiv \alpha \mod 2n', \alpha \in S$ and $\widehat{u}_i$ is chosen uniformly at random from $\mathbb{Z}_q$, otherwise.

When $S = \{\alpha\}$ consists of a single element, we abuse notation and write the Leaky-RLWE parameters as (n', α). Due to automorphisms on the NTT transform, Leaky-RLWE with parameters (n', S) where $S = \{\alpha_1, \alpha_2, \ldots, \alpha_t\}$, is equivalent to Leaky-RLWE with parameters (n', S'), where $S' = \alpha_1^{-1} \cdot S$ (multiply every element of S by α_1^{-1}).

1.2 Our Results

We show the following:

Theorem 1.1 (Informal) *Assuming that Leaky-DRLWE with leakage parameters $(8, \alpha = 1)$ and RLWE parameters $n = 1024, q = 12,289$ and error distribution χ^1 is hard, the shared key v (which is then hashed using a random oracle) of*

---

[1] χ is a rounded Gaussian with standard deviation $\sqrt{8}$, as in the NewHope.

the NewHope key exchange protocol is computationally indistinguishable from a random variable with average min-entropy 237, conditioned on the transcript and leakage of $[\hat{s}, \hat{e}, \hat{s}', \hat{e}', \hat{e}'']_{i \equiv \alpha \mod 16}$.

Moreover, using known relationships between average min-entropy and min-entropy, we have that with all but 2^{-80} probability, the shared key v is indistinguishable from a random variable that has min-entropy 157, conditioned on the transcript and leakage. Note that without leakage, the min-entropy is only 256. This means that the number of leaked bits is far larger than the min-entropy, so the fact that any entropy remains is non-trivial. Indeed, bounding the remaining entropy will require precise analysis of the "key reconciliation" step of the NewHope algorithm.

As mentioned above, due to automorphisms on the NTT transform [38], setting $\alpha = 1$ is WLOG, and α can be any value in $\mathbb{Z}_{16}^*$. While the above may seem straightforward, given that we are already assuming hardness of Leaky-DRLWE, the challenge comes not in the computational part of the analysis (which indeed essentially substitutes instances of Leaky-DRLWE for instances of DRLWE), but in the information-theoretic part of the analysis. Specifically, we must show that given the adversary's additional knowledge about v, as well as the transcript, which includes the reconciliation information (corresponding to the output of a randomized function of v), the input v to the random oracle still has sufficiently high min-entropy. For a discussion of our proof techniques, see Sect. 1.3.

The above theorem could be made more general, and stated in asymptotic form for broader settings of leakage parameters (n', S). However, there is one step in the proof that is not fully generic (although we believe it should hold for wide ranges of parameters) and so for simplicity we choose to state the theorem in terms of the concrete parameters above. Very informally, for the proof to go through, we need to argue existence of a vector of a certain form, where existence depends on the parameter settings of n, q, n' and S. For this step of the proof we can apply a heuristic argument and we confirm existence experimentally for the concrete parameter settings.

We discuss the details of the heuristic argument in Sect. 4.4.

Choice of $n' = 8$ ***in Theorem 1.1*** Experimental results from prior work indicated that the search version of Leaky RLWE is easy for parameters $(n', \alpha = 1)$, where $n' = 4$ (recall that setting $\alpha = 1$ is WLOG), and seems hard for parameters $(n', \alpha = 1)$, where $n' = 8$ and $\alpha \in \mathbb{Z}_{16}^*$. This, combined with their search-to-decision reduction, support the conjecture that the decision version of Leaky RLWE holds (i.e. $D_{real,8,1} \approx D_{sim,8,1}$), for the NewHope parameter settings of $n = 1024$, $q = 12289$, and $\chi = \Psi_{16}$, where Ψ_{16} is centered binomial distribution with parameter 16.[2]

---

[2]The centered binomial distribution is defined in Sect. 4.1.

1.3 Technical Overview

1.3.1 Overview of NewHope Algorithm

We start with an overview of the NewHope key-exchange protocol of [3] and then provide the necessary details relevant to this work. The protocol starts by server P_1 choosing a uniform random polynomial from ring R_q as public key a (note that the elements of R_q are polynomials) and sharing it with client P_2. Both P_1 and P_2 sample the RLWE secrets (resp. errors) s and s' locally. The parties then exchange the RLWE samples b, u.

At this point both the parties share an approximate of shared secret $a \cdot s \cdot s'$. P_2 then generates some additional information r using P_1's RLWE instance b, and shares it with P_1. Both the parties then apply a reconciliation function Rec on their approximate inputs locally. The protocol ensures that after running Rec, the parties agree on the exact same value v.

Finally, the parties apply hash function on v (as instantiation of random oracle) to agree on the key. Thus, the security proof can now rely on the unpredictability of random oracle on input v, rather than arguing that v is indistinguishable from a uniform random value.

1.3.2 Resilience of NewHope to Partial Key Exposure

Recall that P_2 generates additional information r for P_1, which is generated by applying a function HelpRec locally on input v derived using P_1's RLWE instance b and P_2's secret s'. The ring element $v \in \mathbb{Z}_q^n$ that is input to the HelpRec function in the NewHope protocol is split into vectors $\mathbf{x}_i \in \mathbb{Z}_q^4$, $i \in \{0, \ldots, n/4 - 1\}$ and then the HelpRec function is run individually on each $\mathbf{x}_i$. It is not hard to show that, under the Leaky-DRLWE assumption, the distribution over the $\mathbf{x}_i$ (given the transcript and the leakage), for $i \in \{n/8, \ldots, n/4 - 1\}$ is indistinguishable from uniform random in $\mathbb{Z}_q^4$ and for $i \in \{0, \ldots, n/8 - 1\}$, is indistinguishable from uniform random, given a single linear constraint. Specifically, for $i \in \{0, \ldots, n/8 - 1\}$, the $\mathbf{x}_i$ is uniform random, conditioned on $c_{\omega,\alpha} \cdot \mathbf{x}_i = \gamma_i$, for a known $c_{\omega,\alpha}$ and γ_i. The technically difficult part of the proof is showing that, with high probability over γ_i, the min-entropy of $\mathsf{Rec}(\mathbf{x}_i, \mathbf{r}_i)$ is close to 1, conditioned on both the output of $\mathsf{HelpRec}(\mathbf{x}_i; b) = \mathbf{r}_i$ (for a bit $b \in \{0, 1\}$) and the linear constraint $c_{\omega,\alpha} \cdot \mathbf{x}_i = \gamma_i$. This indicates that the probability of guessing the corresponding bit is close to $1/2$, even with respect to an adversary who sees both the transcript and the leakage.

We handle this by showing the existence of a bijective map: $(\mathbf{x}_i, b') \rightarrow (\mathbf{x}'_i, b' \oplus 1)$ such that, $\mathsf{HelpRec}(\mathbf{x}_i, b) = \mathsf{HelpRec}(\mathbf{x}'_i, b') (= \mathbf{r})$ with high probability $1 - p$, and it guarantees $\mathsf{Rec}(\mathbf{x}_i, \mathbf{r}) = 1 \oplus \mathsf{Rec}(\mathbf{x}'_i, \mathbf{r})$. Specifically, we set $\mathbf{x}' = \mathbf{x} + \mathbf{w}$ as the bijective relation. Unlike the original proof from NewHope protocol where $\mathbf{w}_i = (b - b' + q)(1/2, 1/2, 1/2, 1/2)$, we need $\mathbf{w}_i$ to be close to $(q/2, q/2, q/2, q/2)$ and also satisfy an additional linear constraint $c_{\omega,\alpha} \cdot \mathbf{w}_i = 0$ to ensure $c_{\omega,\alpha} \cdot \mathbf{x}'_i = \gamma_i$, which is the information that can be derived about $\mathbf{x}_i$ for $i \in \{0, \ldots, n/8 - 1\}$ from the leakage. In this setting, we can easily prove that if $\mathsf{HelpRec}(\mathbf{x}_i, b) = (\mathbf{x}_i, b)$ $(= \mathbf{r})$ then $\mathsf{Rec}(\mathbf{x}_i, \mathbf{r}) = 1 \oplus \mathsf{Rec}(\mathbf{x}'_i, \mathbf{r})$ following similar argument as in NewHope paper. Then it remains to show that $\mathsf{HelpRec}(\mathbf{x}_i, b) = \mathsf{HelpRec}(\mathbf{x}'_i, b')$ $(= \mathbf{r})$ with high probability $1 - p$. Since $\mathsf{HelpRec}(\mathbf{x}; b) = \mathsf{CVP}_{\tilde{D}_4}\left(\frac{2^r}{q}(\mathbf{x} + b\mathbf{g})\right) \mod 2^r$ as defined, it is equivalent to prove $\mathsf{CVP}_{\tilde{D}_4}(\mathbf{z}) = \mathsf{CVP}_{\tilde{D}_4}(\mathbf{z} + \boldsymbol{\beta})$ with high probability $1 - p$, where $\mathbf{z}, \boldsymbol{\beta}$ are variables that depend on $\mathbf{x}, \mathbf{w}$ which are later defined explicitly in Sect. 4.1.[3] We then analyze the case-by-case probability that algorithm $\mathsf{CVP}_{\tilde{D}_4}$ on input $\mathbf{z}$ and on input $\mathbf{z} + \boldsymbol{\beta}$ disagree in the first three steps and eventually bound the probability that $\mathsf{CVP}_{\tilde{D}_4}(\mathbf{z}) \neq \mathsf{CVP}_{\tilde{D}_4}(\mathbf{z} + \boldsymbol{\beta})$.

1.4 Related Work

Partial Key Exposure There is a large body of work on partial key exposure attacks on RSA, beginning with the seminal work of Boneh et al. [10]. Partial key exposure attacks on RSA are based on a cryptanalytic method known as Coppersmith's method [18, 19]. There has been a long sequence of improved partial key exposure attacks on RSA, see for example [8, 30, 48, 50].

Leakage-Resilient Cryptography The study of provably secure, leakage-resilient cryptography was introduced by the work of Dziembowski and Pietrzak in [29]. Pietrzak [46] also constructed a leakage-resilient stream-cipher. Brakerski et al. [16] showed how to construct a schemes secure against an attacker who leaks at each time period. There are other works as well considering continual leakage [26, 36]. There are also work on leakage-resilient signature scheme [13, 35, 40].

Robustness of Lattice-Based Scheme One of the first and important work is by Goldwasser et al. [33] which shows that LWE is secure even in the cases where secret key is taken from an arbitrary distribution with sufficient entropy and even in the presence of hard-to-invert auxiliary inputs. Additionally they constructed a symmetric-key encryption scheme based on standard LWE assumption, that is robust to secret key leakage. Authors of [1] showed that the public-key scheme of [47] is robust against an attacker which can measure large fraction of secret key without increasing the size of secret key. Dodis et al. [27] presented construction in the case where the leakage is a one way function of the secret (exponentially hard to invert). Their construction are related to LWE assumptions. Dodis et al. [25] presented a construction of public-key cryptosystems based on LWE in the case where the adversary is given any computationally uninvertible function of the secret key. Albrecht et al. [2] consider the ring-LWE and investigate cold boot

---

[3]Here, $\mathbf{v}' = \mathsf{CVP}_{\mathcal{L}}(\mathbf{v})$ denotes, the output of running Closest Vector Problem solver, on input vector $\mathbf{v}$, which returns vector $\mathbf{v}'$ which is the closest lattice vector to $\mathbf{v}$ in lattice $\mathcal{L}$. The lattice $\tilde{D}_4$, is defined in Sect. 4.1.

attacks on schemes based on these problem. They specifically looked into two representation of secret key, namely, polynomial coefficients and encoding of the secret key using a number theoretic transform (NTT). Dachman-Soled et al. [20] considered the leakage resilience of a RLWE-based public key encryption scheme for specific leakage profiles. Stange [49] is showed that given multiple samples of RLWE instances such that the public key for every instance lies in some specific subring, one can reduce the original RLWE problem to multiple independent RLWE problems over the subring.

Recently, Bolboceanu et al. [9] considered the setting of "Order LWE," where the LWE secret is sampled from an *order*. One example of this considered by [9] is sampling the RLWE secret from an ideal $I \subseteq qR$. It is straightforward to see that sampling the RLWE secret uniformly at random from R_q and then leaking the NTT coordinates i such that $i = \alpha \mod 2n'$ is *equivalent* to sampling the RLWE secret from the ideal I that contains it those elements whose NTT transform is 0 in positions i such that $i = \alpha \mod 2n'$. Bolboceanu et al. [9] present both positive results (cases in which a security reduction can still be proved) and a negative result (cases in which the LWE assumption can be broken). However, when fixes the leakage rate to $1/8$, neither of these results covers practical parameter ranges of dimension, modulus and noise rate. As mentioned previously, their decisional assumption is strictly stronger than the assumption adopted in this work. In recent work of Brakerski and Döttling [15] it was shown that the distributions with sufficiently high noise lossiness will lead to hard instances of entropic LWE, which are special kind of LWE samples where the distribution of the secret can be from a family of distributions.

Lattice-Based Key Exchange An important research direction is the design of practical, lattice-based key exchange protocols, which are post-quantum secure. Some of the most influential proposed key exchange protocols include those introduced by Ding [24], Peikert [43], and the NewHope protocol of Alkim et al. [4]. We also mention the Frodo protocol of Bos et al. [11], the Kyber protocol of Bos et al [12], the NTRU protocol of Chen at al. [17], the Round5 protocol of Garcia-Morchon et al. [31], the SABER protocol of D'Anvers et al. [23], the Threebears protocols of Hamburg [34], the NTRU Prime protocol of Bernstein et al. [7], and the LAC protocol of Lu et al. [37], which are the other lattice-based KEMs that were selected as candidates for Round 2 of the NIST Post-Quantum Cryptography standardization effort.

2 Preliminaries

For a positive integer n, we denote by $[n]$ the set $\{0, \ldots, n-1\}$. We denote vectors in boldface x and matrices using capital letters A. For vector x over $\mathbb{R}^n$ or $\mathbb{C}^n$, define the ℓ_2 norm as $\|x\|_2 = \left(\sum_i |x_i|^2 \right)^{1/2}$. We write as $\|x\|$ for simplicity. We use the notation $\approx_{t(n),p(n)}$ to indicate that adversaries running in time $t(n)$ can distinguish two distributions with probability at most $p(n)$.

2.1 Lattices and Background

Let $\mathbb{T} = \mathbb{R}/\mathbb{Z}$ denote the cycle, i.e. the additive group of reals modulo 1. We also denote by $\mathbb{T}_q$ its cyclic subgroup of order q, i.e., the subgroup given by $\{0, 1/q, \ldots, (q-1)/q\}$.

Let H be a subspace, defined as $H \subseteq \mathbb{C}^{\mathbb{Z}_m^*}$, (for some integer $m \geq 2$),

$$H = \{x \in \mathbb{C}^{\mathbb{Z}_m^*} : x_i = \overline{x_{m-i}}, \forall i \in \mathbb{Z}_m^*\}.$$

A *lattice* is a discrete additive subgroup of H. We exclusively consider the full-rank lattices, which are generated as the set of all linear integer combinations of some set of n linearly independent *basis* vectors $B = \{b_j\} \subset H$:

$$\Lambda = \mathcal{L}(B) = \left\{ \sum_j z_j b_j : z_j \in \mathbb{Z} \right\}.$$

The *determinant* of a lattice $\mathcal{L}(B)$ is defined as $|\det(B)|$, which is independent of the choice of basis B. The *minimum distance* $\lambda_1(\Lambda)$ of a lattice Λ (in the Euclidean norm) is the length of a shortest nonzero lattice vector.

The *dual lattice* of $\Lambda \subset H$ is defined as following, where $\langle \cdot, \cdot \rangle$ denotes the inner product.

$$\Lambda^\vee = \{y \in H : \forall x \in \Lambda, \langle x, \overline{y} \rangle = \sum_i x_i y_i \in \mathbb{Z}\}.$$

Note that, $(\Lambda^\vee)^\vee = \Lambda$, and $\det(\Lambda^\vee) = 1/\det(\Lambda)$.

Theorem 2.1 *Let $\mathcal{L} \subseteq \mathbb{R}^n$ be a full dimensional lattice, and let B denote a basis of $\mathcal{L}$. Let $K \subseteq \mathbb{R}^n$ be a convex body. Let $\varepsilon > 0$ denote a scaling such that $\mathcal{P}(B) \cup -\mathcal{P}(B) \subseteq \varepsilon K$. For all $r > \varepsilon$, we have that*

$$(r - \varepsilon)^n \frac{\text{Vol}_n(K)}{\det(\mathcal{L})} \leq |rK \cap \mathcal{L}| \leq (r + \varepsilon)^n \frac{\text{Vol}_n(K)}{\det(\mathcal{L})}.$$

Proof Details can be found in [22]. $\qquad\qquad\qquad\qquad\qquad\qquad\qquad\qquad\square$

2.2 Volume of Hypercube Clipped by One Hyperplane

In this subsection, we consider a unit hypercube and a half hyperspace over n-dimension and want to know volume of their intersection, which can be handled by the following theorem.

Let $[n]$ be an ordered set $\{0, 1, \ldots, n - 1\}$. Let $|\cdot|$ denote the cardinality of a set. For $\mathbf{v} = (v_0, v_1, \ldots, v_{n-1}) \in \mathbb{R}^n$, we define $\mathbf{v_0}$ as $\mathbf{v_0} := \{i \in [n] \mid v_i = 0\}$. Let F^0 be a set of all vertices that each coordinate is either 0 or 1, written as $F^0 = \{(v_0, v_1, \ldots, v_{n-1}) \mid v_i = 0 \text{ or } 1 \text{ for all } i.\}$.

Theorem 2.2 ([6], Revisited by [41, Theorem 1])

$$\text{vol}([0, 1]^n \cap H^+) = \sum_{\mathbf{v} \in F^0 \cap H^+} \frac{(-1)^{|\mathbf{v_0}|} g(\mathbf{v})^n}{n! \prod_{t=1}^n a_t},$$

where the half space H_1^+ is defined by

$$\{\mathbf{t} \mid g(\mathbf{t}) := \mathbf{a} \cdot \mathbf{t} + r_1 = a_0 x_0 + a_1 x_1 + \cdots + a_{n-1} x_{n-1} + r_1 \geq 0\}$$

with $\prod_{t=1}^n a_t \neq 0$.

We now present some background on Algebraic Number Theory.

2.3 Algebraic Number Theory

For a positive integer m, the mth *cyclotomic number field* is a field extension $K = \mathbb{Q}(\zeta_m)$ obtained by adjoining an element ζ_m of order m (i.e. a primitive mth root of unity) to the rationals. The minimal polynomial of ζ_m is the mth *cyclotomic polynomial*

$$\Phi_m(X) = \prod_{i \in \mathbb{Z}_m^*} (X - \omega_m^i) \in \mathbb{Z}[X],$$

where $\omega_m \in \mathbb{C}$ is any primitive mth root of unity in $\mathbb{C}$.

For every $i \in \mathbb{Z}_m^*$, there is an embedding $\sigma_i : K \to \mathbb{C}$, defined as $\sigma_i(\zeta_m) = \omega_m^i$. Let $n = \varphi(m)$, the totient of m. The *trace* $\text{Tr} : K \to \mathbb{Q}$ and *norm* $N : K \to \mathbb{Q}$ can be defined as the sum and product, respectively, of the embeddings:

$$\text{Tr}(x) = \sum_{i \in [n]} \sigma_i(x) \text{ and } N(x) = \prod_{i \in [n]} \sigma_i(x).$$

For any $x \in K$, the l_p *norm* of x is defined as $\|x\|_p = \|\sigma(x)\|_p = (\sum_{i \in [n]} |\sigma_i(x)|^p)^{1/p}$. We omit p when $p = 2$. Note that the appropriate notion of norm $\|\cdot\|$ is used throughout this paper depending on whether the argument is a vector over $\mathbb{C}^n$, or whether the argument is an element from K; whenever the context is clear.

2.4 Ring of Integers and Its Ideals

Let $R \subset K$ denote the set of all algebraic integers in a number field K. This set forms a ring (under the usual addition and multiplication operations in K), called the *ring of integers* of K. Ring of integers in K is written as $R = \mathbb{Z}[\zeta_m]$.

The (absolute) discriminant Δ_K of K measures the geometric sparsity of its ring of integers. The discriminant of the mth cyclotomic number field K is

$$\Delta_K = \left(\frac{m}{\prod_{\text{prime } p \mid m} p^{1/(p-1)}} \right)^n \leq n^n,$$

in which the product in denominator runs over all the primes dividing m.

An *(integral) ideal* $\mathcal{I} \subseteq R$ is a non-trivial (i.e. $\mathcal{I} \neq \varnothing$ and $\mathcal{I} \neq \{0\}$) additive subgroup that is closed under multiplication by R, i.e., $r \cdot a \in \mathcal{I}$ for any $r \in R$ and $a \in \mathcal{I}$. The *norm* of an ideal $\mathcal{I} \subseteq R$ is the number of cosets of $\mathcal{I}$ as an addictive subgroup in R, defined as *index* of $\mathcal{I}$, i.e., $\mathrm{N}(\mathcal{I}) = |R/\mathcal{I}|$. Note that $\mathrm{N}(\mathcal{I}\mathcal{J}) = \mathrm{N}(\mathcal{I})\mathrm{N}(\mathcal{J})$.

A *fractional* ideal $\mathcal{I}$ in K is defined as a subset such that $\mathcal{I} \subseteq R$ is an integral ideal for some nonzero $d \in R$. Its norm is defined as $\mathrm{N}(\mathcal{I}) = \mathrm{N}(d\mathcal{I})/\mathrm{N}(d)$. An *ideal lattice* is a lattice $\sigma(\mathcal{I})$ embedded from a fractional ideal $\mathcal{I}$ by σ in H. The determinant of an ideal lattice $\sigma(\mathcal{I})$ is $\det(\sigma(\mathcal{I})) = \mathrm{N}(\mathcal{I}) \cdot \sqrt{\Delta_K}$. For simplicity, however, most often when discussing about ideal lattice, we omit mention of σ since no confusion is likely to arise.

Lemma 2.3 ([39]) *For any fractional ideal $\mathcal{I}$ in a number field K of degree n,*

$$\sqrt{n} \cdot \mathrm{N}^{1/n}(\mathcal{I}) \leq \lambda_1(\mathcal{I}) \leq \sqrt{n} \cdot \mathrm{N}^{1/n}(\mathcal{I}) \cdot \sqrt{\Delta_K^{1/n}}.$$

For any *fractional* ideal $\mathcal{I}$ in K, its *dual* ideal is defined as

$$\mathcal{I}^\vee = \{a \in K : \mathrm{Tr}(a\mathcal{I}) \subset \mathbb{Z}\}.$$

Definition 2.4 For $R = \mathbb{Z}[\zeta_m]$, define $g = \prod_p (1 - \zeta_p) \in R$, where p runs over all odd primes dividing m. Also, define $t = \frac{\hat{m}}{g} \in R$, where $\hat{m} = \frac{m}{2}$ if m is even, otherwise $\hat{m} = m$.

The dual ideal $R^\vee$ of R is defined as $R^\vee = \langle t^{-1} \rangle$, satisfying $R \subseteq R^\vee \subseteq \hat{m}^{-1} R$. For any fractional ideal $\mathcal{I}$, its dual is $\mathcal{I}^\vee = \mathcal{I}^{-1} \cdot R^\vee$. The quotient $R_q^\vee$ is defined as $R_q^\vee = R^\vee / q R^\vee$.

Fact 2.5 ([39]) *Assume that q is a prime satisfying $q = 1 \mod m$, so that $\langle q \rangle$ splits completely into n distinct ideals of norm q. The prime ideal factors of $\langle q \rangle$ are $\mathfrak{q}_i = \langle q \rangle + \langle \zeta_m - \omega_m^i \rangle$, for $i \in \mathbb{Z}_m^*$. By Chinese Reminder Theorem, the natural ring homomorphism $R/\langle q \rangle \to \prod_{i \in \mathbb{Z}_m^*} (R/\mathfrak{q}_i) \cong (\mathbb{Z}_q^n)$ is an isomorphism.*

2.5 Ring-LWE

We next present the formal definition of the ring-LWE problem as given in [39].

Definition 2.6 (Ring-LWE Distribution) For a "secret" $s \in R_q^\vee$ (or just $R^\vee$) and a distribution χ over $K_\mathbb{R}$, a sample from the ring-LWE distribution $A_{s,\chi}$ over $R_q \times (K_\mathbb{R}/q R^\vee)$ is generated by choosing $a \leftarrow R_q$ uniformly at random, choosing $e \leftarrow \chi$, and outputting $(a, b = a \cdot s + e \mod q R^\vee)$.

Definition 2.7 (Ring-LWE, Average-Case Decision) The average-case decision version of the ring-LWE problem, denoted $\text{DRLWE}_{q,\chi}$, is to distinguish with non-negligible advantage between independent samples from $A_{s,\chi}$, where $s \leftarrow \chi$ is sampled from the error distribution, and the same number of uniformly random and independent samples from $R_q \times (K_\mathbb{R}/q R^\vee)$.

Theorem 2.8 ([39, Theorem 2.22]) *Let K be the mth cyclotomic number field having dimension $n = \varphi(m)$ and $R = \mathcal{O}_K$ be its ring of integers. Let $\alpha = \alpha(n) > 0$, and $q = q(n) \geq 2$, $q = 1 \mod m$ be a $\text{poly}(n)$-bounded prime such that $\alpha q \geq \omega(\sqrt{\log n})$. Then there is a polynomial-time quantum reduction from $\tilde{O}(\sqrt{n}/\alpha)$-approximate SIVP (or SVP) on ideal lattices in K to the problem of solving $\text{DRLWE}_{q,\chi}$ given only l samples, where χ is the Gaussian distribution D_ξ for $\xi = \alpha \cdot q \cdot (nl/\log(nl))^{1/4}$.*

2.6 Number Theoretic Transform (NTT)

Let $R_q := \mathbb{Z}_q[x]/x^n + 1$ be the ring of polynomials, with $n = 2^d$ for any positive integer d. Also, let $m = 2n$ and $q = 1 \mod m$. For, ω a mth root of unity in $\mathbb{Z}_q$ the NTT of polynomial $p = \sum_{i=0}^{n-1} p_i x^i \in R_q$ is define as,

$$\hat{p} = \text{NTT}(p) := \sum_{i=0}^{n-1} \hat{p}_i x^i$$

where the NTT coefficients $\widehat{p}_i$ are defined as: $\widehat{p}_i = \sum_{j=0}^{n-1} p_j \omega^{j(2i+1)}$.

The function NTT^{-1} is the inverse of function NTT, defined as

$$p = \mathsf{NTT}^{-1}(\widehat{p}) := \sum_{i=0}^{n-1} p_i x^i$$

where the NTT inverse coefficients p_i are defined as: $p_i = n^{-1} \sum_{j=0}^{n-1} \widehat{p}_j \omega^{i(2j+1)}$.

We next present the definitions of min-entropy and average min-entropy.

2.7 Min-Entropy and Average Min-Entropy

Definition 2.9 (Min-Entropy) A random variable X has *min-entropy* k, denoted $H_\infty(X) = k$, if

$$\max_x \Pr[X = x] = 2^{-k}.$$

Definition 2.10 (Average Min-Entropy) Let (X, Z) be a pair of random variables. The *average min entropy* of X conditioned on Z is

$$\tilde{H}_\infty(X \mid Z) \stackrel{\text{def}}{=} -\log E_{z \leftarrow Z} \max_x \Pr[X = x \mid Z = z].$$

Lemma 2.11 ([28]) *For any* $\delta > 0$, $H_\infty(X \mid Z = z)$ *is at least* $\tilde{H}_\infty(X \mid Z) - \log(1/\delta)$ *with probability at least* $1 - \delta$ *over the choice of* z.

3 Search and Decisional RLWE with Leakage

In this section we define the search and decisional ring-LWE problem with structured leakage on the secret key (i.e. partial key exposure). The definition is similar to the Definition 2.7.

Ring elements (polynomials) p are stored as a vector of their coefficients $(p_0, \ldots, p_{n-1})$. For $p \in R_q$ we denote $\widehat{p} := \mathsf{NTT}(p) := (p(\omega^1), p(\omega^3), \ldots, p(\omega^{2n-1}))$, where ω is a $2n$-th primitive root of unity in $\mathbb{Z}_q$ (which exists since q is prime and $q \equiv 1 \mod 2n$), and $p(\omega^i)$ for $i \in \mathbb{Z}_{2n}^*$ denotes evaluation of the polynomial p at ω^i. Note that $\widehat{p}$ is indexed by the set $\mathbb{Z}_{2n}^*$.

Definition 3.1 (Ring-LWE, Search with Leakage) The search version of the ring-LWE problem with leakage, denoted $\mathsf{SRLWE}_{q,\psi,n',\mathcal{S}}$, is parameterized by $(n' \in \{1, 2, 4, 8, \ldots n\}, \mathcal{S} \subseteq \mathbb{Z}_{2n'}^*)$. The experiment chooses $s \leftarrow \chi$, where $s = \mathsf{NTT}^{-1}(\widehat{s})$. The goal of the adversary is to recover s, given independent samples from the distribution $D_{real,n',S}$, which outputs $\left(\widehat{a}, \widehat{a} \cdot \widehat{s} + \widehat{e}, [\widehat{s_i}]_{i \equiv \alpha \bmod 2n'} |_{\forall \alpha \in S}\right)$ where a, e are obtained from $A_{s,\psi}$ as described in Definition 2.6.

Definition 3.2 (Ring-LWE, Decision with Leakage) The decision version of the ring-LWE problem with leakage, denoted Leaky-DRLWE$_{q,\psi,n',S}$, is parameterized by $(n' \in \{1, 2, 4, 8, \ldots n\}, S \subseteq \mathbb{Z}^*_{2n'})$. The experiment chooses $s \leftarrow \chi$, where $s = \text{NTT}^{-1}(\widehat{s})$. The goal of the adversary is to distinguish between independent samples from the distributions $D_{real,n',S}$ and $D_{sim,n',S}$, where $D_{real,n',S}$ outputs $\left(\widehat{a}, \widehat{a} \cdot \widehat{s} + \widehat{e}, [\widehat{s_i}]_{i \equiv \alpha \bmod 2n'} |_{\forall \alpha \in S}\right)$ where a, e are obtained from $A_{s,\psi}$ as described in Definition 2.6. And the $D_{sim,n',S}$ outputs $\left(\widehat{a}, \widehat{u}, [\widehat{s_i}]_{i \equiv \alpha \bmod 2n'} |_{\forall \alpha \in S}\right)$ where a, e are obtained from $A_{s,\psi}$ as described in Definition 2.6, and

$$\widehat{u_i} = \widehat{a_i} \cdot \widehat{s_i} + \widehat{e_i} \quad | \quad i \equiv \alpha \bmod 2n' \, \forall \alpha \in S$$

and

$$\widehat{u_i} \leftarrow \mathbb{Z}_q$$

chosen uniformly random, otherwise.

Note that in the above definitions, the adversary can receive the leakage $[\widehat{e_i}]_{i \equiv \alpha \bmod 2n'} |_{\forall \alpha \in S}$ for each error vector as well, since given $\widehat{a}$ and $[\widehat{s_i}]_{i \equiv \alpha \bmod 2n'} |_{\forall \alpha \in S}$, the adversary can derive $[\widehat{e_i}]_{i \equiv \alpha \bmod 2n'} |_{\forall \alpha \in S}$.

Fact 3.3 *If decisional RLWE with leakage parameterized by (n', S) as above is hard for uniformly distributed $\widehat{a}$, then it is also hard for $\widehat{a}$ that is arbitrarily distributed in positions i such that $i \equiv \alpha \mod 2n', \alpha \in S$ and uniformly distributed elsewhere.*

This is because given an RLWE instance with leakage $\left(\widehat{a}, \widehat{u}, [\widehat{s_i}]_{i \equiv \alpha \bmod 2n'} |_{\forall \alpha \in S}\right)$, for $i \equiv \alpha \mod 2n', \alpha \in S$ one can change the instance from $\widehat{a_i}$ to $\widehat{a_i'}$ by adding $(\widehat{a_i'} - \widehat{a_i}) \cdot \widehat{s_i}$ from the i-th coordinate of $\widehat{u}$.

When $S = \{\alpha\}$ consists of a single element, we abuse notation and write the Leaky-RLWE parameters as (n', α).

4 Leakage Analysis of New Hope Key Exchange

4.1 New Hope Key Exchange Scheme

It contains New Hope key exchange scheme and subroutines of HelpRec and Rec. In this section we revise some important results and algorithms from [3].

Let $\tilde{D}_4$ be a lattice as defined below:

$$\tilde{D}_4 = \mathbb{Z}^4 \cup g + \mathbb{Z}^4 \text{ where } g^t = \left(\frac{1}{2}, \frac{1}{2}, \frac{1}{2}, \frac{1}{2}\right)$$

Let, $B = (u_0, u_1, u_2, g)$ be the basis of $\tilde{D}_4$, where u_i are the canonical basis vectors of $\mathbb{Z}^4$. Note that $u_3 = B \cdot (-1, -1, -1, 2)^t$. Also, let $\mathcal{V}$ be the Voronoi cell of $\tilde{D}_4$.[4]

Note that, u_0, u_1, u_2, and $2g$ are in $\mathbb{Z}^4$. Therefore, a vector in $\tilde{D}_4/\mathbb{Z}^4$ can be checked by simply checking the parity of its last coordinate when represented with basis B. We can now use a simple encoding and decoding scheme to represent a bit. The encoding algorithm is as follows: $\mathsf{Encode}(k \in \{0, 1\}) = kg$. For decoding to $\tilde{D}_4/\mathbb{Z}^4$, the correctness requires that the error vector $e \in \mathcal{V}$. As noted in [3], this is equivalent to checking if $\|e\|_1 \leq 1$. We can now present the decoding algorithm as follows in Algorithm 4.1:

Algorithm 4.1 (Algorithm 1) Decode $(x \in \mathbb{R}^4/\mathbb{Z}^4)$

Ensure : A bit k such that kg is closest vector to $x + \mathbb{Z}^4 : x - kg \in \mathcal{V} + \mathbb{Z}^4$

1. $v = x - \lfloor x \rceil$
2. **return** 0 if $\|v\|_1 \leq 1$ and 1 otherwise

Lemma 4.2 (Lemma C.1 [3]) *For any $k \in \{0, 1\}$ and any $e \in \mathbb{R}^4$ such that $\|e\|_1 < 1$, we have $\mathsf{Decode}(kg + e) = k$.*

Let us now present the algorithm CVP (Closest Vector Problem), which will be used as subroutine in reconciliation algorithms, as follows:

Algorithm 4.3 (Algorithm 2) $\mathsf{CVP}_{\tilde{D}_4} (x \in \mathbb{R}^4)$

Ensure : An integer vector z such that Bz is closest vector to $x : x - Bz \in \mathcal{V}$
1. $v_0 \leftarrow \lfloor x \rceil$
2. $v_1 \leftarrow \lfloor x - g \rceil$
3. **if** $(\|x - v_0\| < 1)$ **then** $k \leftarrow 0$ **else** $k \leftarrow 1$
4. $(v_0, v_1, v_2, v_3)^t \leftarrow v_k$
5. **return** $(v_0, v_1, v_2, k)^t + v_3 \cdot (-1, -1, -1, 2)^t$

---

[4]For more details and background on reconciliation mechanism of NewHope, please refer to [3] (section 5 and appendix C).

Next, we define the r-bit reconciliation as,

$$\mathsf{HelpRec}(x; b) = \mathsf{CVP}_{\tilde{D}_4}\left(\frac{2^r}{q}(x + bg)\right) \bmod 2^r,$$

where $b \in \{0, 1\}$ is a uniformly chosen random bit.

Lemma 4.4 (Lemma C.2 [3]) *Assume* $r \geq 1$ *and* $q \geq 9$. *For any* $x \in \mathbb{Z}_q^4$, *set* $r := \mathsf{HelpRec}(x) \in \mathbb{Z}_{2^r}^4$. *Then,* $\frac{1}{q}x - \frac{1}{2^r}Br \bmod 1$ *is close to a point of* $\tilde{D}_4/\mathbb{Z}^4$, *precisely, for* $\alpha = \frac{1}{2^r} + \frac{2}{q}$: $\frac{1}{q}x - \frac{1}{2^r}Br \in \alpha\mathcal{V} + \mathbb{Z}^4$ *or* $\frac{1}{q}x - \frac{1}{2^r}Br \in g + \alpha\mathcal{V} + \mathbb{Z}^4$. *Additionally, for* x *uniformly chosen in* $\mathbb{Z}_q^4$ *we have* $\mathsf{Decode}\left(\frac{1}{q}x - \frac{1}{2^r}Br\right)$ *is uniform in* $\{0, 1\}$ *and independent of* r.

Let, $\mathsf{Rec}(x, r) = \mathsf{Decode}\left(\frac{1}{q}x - \frac{1}{2^r}Br\right)$.
We can now define the following reconciliation protocol:

Algorithm 4.5 (Protocol 1) *Reconciliation protocol in* $q\tilde{D}_4/q\mathbb{Z}^4$

Alice		Bob
$x' \in \mathbb{Z}_q^4$	$x' \approx x$	$x' \in \mathbb{Z}_q^4$
	$\xleftarrow{r}$ $r \leftarrow \mathsf{HelpRec}(x) \in \mathbb{Z}_{2^r}^4$	
$k' \leftarrow \mathsf{Rec}(x', r)$		$k \leftarrow \mathsf{Rec}(x, r)$

Lemma 4.6 (Lemma C.3 [3]) *If* $\|x - x'\|_1 < \left(1 - \frac{1}{2^r}\right) \cdot q - 2$, *then by the above Protocol 4.5* $k = k'$. *Additionally, if* x *is uniform, then* k *is uniform independently of* r.

We define centered binomial distribution Ψ_{16} as the subtraction of two binomial distribution $B(16, 0.5)$ and Ψ_{16}^n is n draw from that distribution.

We now present the complete NewHope key exchange protocol given in [3] as Protocol 4.7.

Algorithm 4.7 (Protocol 2) *Parameters: $q = 12289$, $n = 1024$ Error Distribution: Ψ_{16}*

Alice(server)   **Bob(client)**

Sample : $a \leftarrow R_q$

$s, e \longleftarrow \Psi_{16}^n$ $s', e', e'' \longleftarrow \Psi_{16}^n$

$\widehat{s} := \mathsf{NTT}(s), \widehat{e} := \mathsf{NTT}(e)$ $\widehat{s}' := \mathsf{NTT}(s'), \widehat{e}' := \mathsf{NTT}(e')$

$\widehat{e}'' := \mathsf{NTT}(e'')$

$\widehat{b} := \widehat{a} \cdot \widehat{s} + \widehat{e}$ $\xrightarrow{\widehat{a}, \widehat{b}}$

$\widehat{u} := \widehat{a} \cdot \widehat{s}' + \widehat{e}'$

$\widehat{v} := \widehat{b} \cdot \widehat{s}' + \widehat{e}''$

$v := \mathsf{NTT}^{-1}(\widehat{v})$

$\xleftarrow{(\widehat{u}, r)}$ $r \leftarrow \mathsf{HelpRec}(v)$

$w := \mathsf{NTT}^{-1}(\widehat{u} \cdot \widehat{s})$ $v \leftarrow \mathsf{Rec}(v, r)$

$v \leftarrow \mathsf{Rec}(w, r)$ $\mu \leftarrow \mathsf{SHA3-256}(v)$

$\mu \leftarrow \mathsf{SHA3-256}(v)$

4.2 Security with Auxiliary Inputs

In this section we consider a modification to Protocol 4.7 in which all binomial random variables are instead drawn from discrete Gaussians with corresponding standard deviation σ. Specifically, we assume that the coefficients of the secret and error vectors are drawn from $D_{\tilde{\sigma}}$ (a discretized, Gaussian distribution with probability mass function proportional to $e^{-\pi x^2/(\tilde{\sigma}^2)}$), where $\tilde{\sigma} = \sqrt{2\pi} \cdot \sigma$.[5]

We prove in Corollary 4.15 that the distribution over v, given the transcript of the modified protocol, is (with all but negligible probability) indistinguishable from a distribution with high min-entropy. By the analysis of [3] leveraging Renyi divergence and the random oracle model, this is sufficient to argue security in the presence of leakage.

The proof of Corollary 4.15 has two components. In the first (computational and information-theoretic) component (proof of Theorem 4.8), we analyze the distribution of v, conditioned on the transcript that *does not include the reconciliation information r* and show that it is close to another distribution over v'. In the second (information-theoretic only) component (proof of Theorem 4.9),

---

[5]The proof of statistical closeness of binomial distribution Ψ_{16}^n, and discrete Gaussian distribution with $\sigma = \sqrt{8}$ can be found in Appendix B of the papers [3, 4] which introduced NewHope protocol.

we analyze the expected min-entropy of $v \leftarrow \mathsf{Rec}(v', r)$, conditioned on the adversary's view which now additionally includes the reconciliation information $r \leftarrow \mathsf{HelpRec}(v', b)$. These are then combined to obtain Corollary 4.15.

The view of the adversary in the modified protocol consists of the tuple

$$\mathsf{View}_A := (\widehat{a}, \widehat{b}, \widehat{u}, [\widehat{s}_i, \widehat{e}_i, \widehat{s}'_i, \widehat{e}'_i, \widehat{e}''_i]_{i \equiv \alpha \mod 2n'}).$$

Moreover, note that $\widehat{v}_i = \widehat{b}_i \cdot \widehat{s}'_i + \widehat{e}''_i$, so $[\widehat{v}_i]_{i \equiv \alpha \mod 2n'}$ is deducible from the view.

Let $\eta_\epsilon(\Lambda)$ be the smoothing parameter of lattice Λ defined as follows:

$$\Lambda := \{\boldsymbol{w} \mid \langle (1, \omega^{n/n'}, \omega^{2n/n'}, \ldots, \omega^{(n'-1)n/n'}), \boldsymbol{w} \rangle = 0 \mod q\},$$

where ω is a $2n$-th primitive root of unity modulo q.

Theorem 4.8 *If the Ring-LWE decision problem with leakage is hard as defined in Sect. 3 with parameters $(n' = 8, \alpha \in \mathbb{Z}^*_{2n'})$, and error distribution $D_{\tilde{\sigma}}$, where $\tilde{\sigma} \geq \eta_\epsilon(\Lambda)$, then*

(1) The marginal distribution over $[\widehat{v}_i]_{i \equiv \alpha \mod 2n'}$, is point-wise, multiplicatively $(1 + \epsilon)^{n/n'}$-close to uniform random over $\mathbb{Z}_q^{n/n'}$, where we may set $\epsilon = 1/166$, when $n' = 8$ for NewHope parameters.

(2) Given the adversary's view, View_A,

$$[\widehat{v}_i]_{i \neq \alpha \mod 2n'}$$

is computationally indistinguishable from uniform random over $\mathbb{Z}_q^{n-n/n'}$.

Proof We prove the above theorem by considering the adversary's view in a sequence of hybrid distributions.

Hybrid H_0 This is the real world distribution

$$(\widehat{a}, \widehat{b}, \widehat{u}, [\widehat{s}_i, \widehat{e}_i, \widehat{s}', \widehat{e}'_i, \widehat{e}''_i]_{i \equiv \alpha \mod 2n'}, \widehat{v}).$$

Hybrid H_1 Here we replace $\tilde{b}$ by $\widehat{b}'$, where $\widehat{b}'_i = \widehat{b}_i$ for $i \equiv \alpha \mod 2n'$ and $\widehat{b}'_i$ is chosen uniformly at random from $\mathbb{Z}_q$ for $i \not\equiv \alpha \mod 2n'$.

$$(\widehat{a}, \widehat{b}', \widehat{u}, [\widehat{s}_i, \widehat{e}_i, \widehat{s}', \widehat{e}'_i, \widehat{e}''_i]_{i \equiv \alpha \mod 2n'}, \widehat{v}).$$

Claim 4.9 $H_0 \approx H_1$

Claim 4.9 follows from the decision ring-LWE with leakage assumption Definition 3.2 and Fact 3.3.

Hybrid H_2 This is same as hybrid H_1 except we replace $\widehat{u}$ by $\widehat{u}'$ and $\widehat{v}$ by $\widehat{v}'$, where $\widehat{u}'_i = \widehat{u}_i, \widehat{v}'_i = \widehat{v}_i$, for $i \equiv \alpha \mod 2n'$ and $\widehat{u}'_i, \widehat{v}'_i$ are chosen uniformly at random from $\mathbb{Z}_q$ for $i \not\equiv \alpha \mod 2n'$.

$$(\widehat{\boldsymbol{a}}, \widehat{\boldsymbol{b}}', \widehat{\boldsymbol{u}}', [\widehat{s}_i, \widehat{e}_i, \widehat{s}', \widehat{e}'_i, \widehat{e}''_i]_{i \equiv \alpha \mod 2n'}, \widehat{\boldsymbol{v}}').$$

Claim 4.10 $H_1 \approx H_2$

Claim 4.10 follows from the decision ring-LWE with leakage assumption Definition 3.2 and Fact 3.3.

We now analyze the distribution over $[\widehat{v}']_{i \equiv \alpha \mod 2n'}$ in Hybrid H_2. First, note that the distribution over $[\widehat{v}']_{i \equiv \alpha \mod 2n'}$ is unchanged in Hybrids H_0, H_1, H_2. Further, we show that for every vector $\boldsymbol{w} \in Z_q^{n/n'}$, $\Pr[[\widehat{v}']_{i \equiv \alpha \mod 2n'} = \boldsymbol{w}] \leq \frac{(1+\epsilon)^{n/n'}}{q^{n/n'}}$ (for ϵ as in the statement of Theorem 4.8). In fact, we will show that the above holds for $[\widehat{e}'']_{i \equiv \alpha \mod 2n'}$ and use the fact that for $i \equiv \alpha \mod 2n'$, $\widehat{v}_i = \widehat{e}''_i + \widehat{b}_i \cdot \widehat{s}_i = \widehat{e}''_i + \widehat{b}_i \cdot \widehat{s}_i$. This is sufficient to prove item (1) of Theorem 4.8.

We observe that there is a bijection between the vector $[\widehat{e}''_i]_{i \equiv \alpha \mod 2n'}$ and the polynomial $\boldsymbol{f} := \boldsymbol{e}'' \mod (x^{n/n'} - (\omega^\alpha)^{n/n'})$, where ω is a $2n$-th primitive root of unity modulo q. We further assume WLOG that $\alpha = 1$, as this does not affect the computations, due to the ring automorphisms [38]. Now, each of the coordinates of $\boldsymbol{f}$ is equal to $f_i := [1, \omega^{n/n'}, \omega^{2n/n'}, \ldots, \omega^{(n'-1)n/n'}] \cdot [\boldsymbol{e}''_{i+j \cdot n/n'}]_{j \in \{0,\ldots,n'-1\}}$, where each coordinate of $\boldsymbol{e}''$ is drawn independently from $D_{\widetilde{\sigma}}$ (a discretized, Gaussian distribution with probability mass function proportional to $e^{-\pi x^2/(\widetilde{\sigma}^2)}$). Since the n/n' coordinates of $\boldsymbol{f}$ are independent, it is sufficient to show that for each coordinate f_i, and any value $\widetilde{f}_i \in Z_q$, $\Pr[f_i = \widetilde{f}_i] \leq \frac{1+\epsilon}{q}$.

Towards this goal, consider the lattice Λ defined as follows:

$$\Lambda := \{\boldsymbol{w} \mid \langle(1, \omega^{n/n'}, \omega^{2n/n'}, \ldots, \omega^{(n'-1)n/n'}), \boldsymbol{w}\rangle = 0 \mod q\}.$$

We will show that $\widetilde{\sigma}$ is larger than the smoothing parameter, $\eta_\epsilon(\Lambda)$, of this lattice. By definition, this implies that for each coordinate f_i, and any value $\widetilde{f}_i \in \mathbb{Z}_q$, $\Pr[f_i = \widetilde{f}_i] \leq \frac{1+\epsilon}{q}$, which completes our argument.

We upperbound $\eta_\epsilon(\Lambda)$, via the bound of [42, Lemma 3.3] on the smoothing parameter of a lattice, observing that Λ has dimension n':

$$\eta_\epsilon(\Lambda) \leq \sqrt{\frac{\ln(2n'(1+1/\epsilon))}{\pi}} \cdot \lambda_{n'}(\Lambda), \tag{1}$$

where $\lambda_{n'}(\Lambda)$ is the n'-th successive minimum of Λ.

Now, we consider applying the above to NewHope parameter settings with $n' = 8$ and $\epsilon = 1/166$. Specifically, with standard deviation $\sigma = \sqrt{8}$ for error (as in NewHope), the discrete Gaussian with the same standard deviation has pdf proportional to

$$e^{-x^2/(2 \cdot \sigma^2)} = e^{-x^2/(2 \cdot \sqrt{8}^2)} = e^{-\pi x^2/(\sqrt{16 \cdot \pi}^2)} = e^{-\pi x^2/(\widetilde{\sigma}^2)},$$

where $\widetilde{\sigma} = 4\sqrt{\pi} \geq 7.0898154$.

Plugging in our parameters in (1), we have that $\eta_{1/166}(\Lambda) \leq \sqrt{\frac{\ln(16 \cdot 167)}{\pi}} \cdot \lambda_8(\Lambda)$, where and $\lambda_8(\Lambda)$ is upperbounded by the maximum length over the vectors in the reduced basis of Λ output by the BKZ algorithm. Since Λ is a dimension 8 lattice, we are able to efficiently compute $\lambda_8(\Lambda) \leq 4.472136$ using Sage's BKZ 2.0 implementation. Thus, we have that $\eta_{1/166}(\Lambda) \leq 7.08753 \leq \tilde{\sigma}$.

We conclude as we have now shown that (1) and (2) of Theorem 4.8 hold in Hybrid 2.   □

Switching from NTT to Polynomial Representation   We showed that in Hybrid H_2, given fixed $[\widehat{v}_i']_{i \equiv \alpha \bmod 2n'}$, the distribution over $[\widehat{v}_i']_{i \not\equiv \alpha \bmod 2n'}$ is uniform random. We now characterize the induced distribution of $\mathbf{x} := \mathbf{v}'$ (i.e. the polynomial form), given $[\widehat{v}_i']_{i \equiv \alpha \bmod 2n'}$. Henceforth, we assume for simplicity that $n' = 8$. Given $[\widehat{v}_i']_{i \equiv \alpha \bmod 16}$ an attacker can recover $\mathbf{y}(x) = \mathbf{v}'(x) \bmod (x^{n/8} - (\omega^\alpha)^{n/8})$. Thus the leaked information forms a linear equation as follow:

$$\sum_{k=0}^{7} (\omega^\alpha)^{\frac{kn}{8}} v'_{\frac{kn}{8}+i} = y_i,$$

where $i \in \{0, \dots, n/8 - 1\}$.

For $i \in \{0, \dots, n/8 - 1\}$, fix $v'_{\frac{kn}{8}+i}$, for $k \in \{1, 3, 5, 7\}$ then we have

$$\sum_{k=0}^{3} (\omega^\alpha)^{\frac{2kn}{n'}} v'_{\frac{2kn}{n'}+i} = y_i - \sum_{\kappa=0}^{3} (\omega^\alpha)^{\frac{(2k+1)n}{8}} v'_{\frac{(2k+1)n}{8}+i}. \tag{2}$$

Let γ_i be the right hand side of (2). Let $\mathbf{c}_{\omega,\alpha} = [1 \quad (\omega^\alpha)^{n/4} \quad (\omega^\alpha)^{n/2} \quad (\omega^\alpha)^{3n/4}]$. Thus the linear constraint corresponding to (2) can be written as $f_{\omega,j}(\mathbf{x}_i) := \mathbf{c}_{\omega,\alpha} \cdot \mathbf{x}_i = \gamma_i$, where $\mathbf{x}_i \in \mathbb{Z}_q^4$. Recall that due to automorphisms, we may assume $\alpha = 1$.

Distributions Over Polynomial Representation   For every fixed setting of $[\widehat{v}_i']_{i \equiv \alpha \bmod 2n'}$, the distribution over $[\mathbf{x}_i]_{i \in \{n/8, \dots, n/4-1\}} = [\mathbf{v}'_{\frac{kn}{8}+j}]_{j \in \{0, \dots, n/8-1\}}$, $k \in \{1,3,5,7\}$ is uniform random. This corresponds to setting $\mathbf{x}_i \leftarrow \mathbb{Z}_q^4$ uniformly at random, for $i \in \{n/8, \dots, n/4 - 1\}$. Given $[\widehat{v}_i']_{i \equiv \alpha \bmod 2n'}$ and the fixed values of $\mathbf{x}_i$, the distribution over $[\mathbf{x}_i]_{i \in \{0, \dots, n/8-1\}} = [\mathbf{v}'_{\frac{kn}{8}+j}]_{j \in \{0, \dots, n/8-1\}, k \in \{0,2,4,6\}}$, which we denote by $\mathcal{S}_\gamma = (\mathcal{S}_{\gamma_0}, \dots, \mathcal{S}_{\gamma_{n/8-1}})$, corresponds to, for each $i \in \{0, \dots, n/8 - 1\}$, choosing $\mathbf{x}_i \in \mathbb{Z}_q^4$ uniformly at random, conditioned on $\mathbf{c}_{\omega,\alpha} \cdot \mathbf{x}_i = \gamma_i$. Let Ψ be the distribution over $\mathbf{y}$.[6] Since (once we fix $[\mathbf{x}_i]_{i \in \{n/8, \dots, n/4-1\}}$) there is a bijection between $[\widehat{v}_i']_{i \equiv \alpha \bmod 2n'}$ and the values of the constraints $[\gamma_i]_{i \in \{0, \dots, n/8-1\}}$, so by (1) of Theorem 4.8 we have that for every fixed $\tilde{\gamma}$, $\Pr_\Psi[\mathbf{y} = \tilde{\gamma}] \leq \frac{(1+1/166)^{128}}{q^{n/8}}$.

---

[6]Not to be confused with Ψ_16 which denotes a centered binomial distribution and was used as an error distribution in Sect. 4.1.

Analyzing the Average Min-Entropy of v To summarize the analysis above, conditioned on the view of the adversary, for each $i \in \{n/8, \ldots, n/4 - 1\}$, $\mathbf{x}_i$ is sampled uniformly and independently. Once these values are fixed, we can consider the resulting distribution Ψ over $\mathbf{y} = \gamma_0, \ldots, \gamma_{n/8-1}$. For each $i \in \{0, \ldots, n/8-1\}$, $\mathbf{x}_i$ is sampled independently from $\mathcal{S}_{\gamma_i}$ (defined in the preceding paragraph).

Clearly, for $i \in \{n/8, \ldots, n/4 - 1\}$, since $\mathbf{x}_i \leftarrow \mathbb{Z}_q^4$ are sampled uniformly at random and independently, we can use the same analysis as in [3] to prove that, conditioned on the output of HelpRec, the output of Rec for $i \in \{n/8, \ldots, n/4 - 1\}$ has (average) min-entropy exactly 1, conditioned on the leakage and transcript. Thus, it remains to show that for $i \in \{0, \ldots, n/8 - 1\}$, conditioned on the output of HelpRec, the output of Rec has high average min-entropy.

For $\gamma_i \in \mathbb{Z}_q$, recall that $\mathcal{S}_{\gamma_i}$ is the set of $\mathbf{x}_i \in \mathbb{Z}_q^4$ that satisfy $\mathbf{c}_{\omega,\alpha} \cdot \mathbf{x}_i = \gamma_i$. Note that the sets $\mathcal{S}_{\gamma_i}, \gamma_i \in \mathbb{Z}_q$ form a partition of $\mathbb{Z}_q^4$. Let $\mathcal{R}_{\gamma_i}$ be the distribution over outputs $\mathbf{r}$ of HelpRec($\mathbf{x}_i; b$) when $\mathbf{x}_i$ is chosen uniformly at random from $\mathcal{S}_{\gamma_i}$ and b is chosen uniformly at random from $\{0, 1\}$.

For $i \in \{0, \ldots, n/8 - 1\}$, the average min-entropy of the output of Rec, conditioned on the output of HelpRec is equal to:

$$
-\log_2 \left(E_{\mathbf{y} \leftarrow \Psi, [\mathbf{r}_i \sim \mathcal{R}_{\gamma_i}]_{i \in \{0, \ldots, n/8-1\}}} \left[\prod \max_{\substack{\beta \in \{0,1\} \ \mathbf{x}_i \sim \mathcal{S}_{\gamma_i} \\ b \sim \{0,1\}}} \Pr \left[\mathsf{Rec}(\mathbf{x}_i, \mathbf{r}_i) = \beta \mid \mathsf{HelpRec}(\mathbf{x}_i; b) = \mathbf{r}_i \right] \right] \right).
$$

We can rewrite the above expression as follows:

$$
E_{\mathbf{y} \leftarrow \Psi, [\mathbf{r}_i \sim \mathcal{R}_{\gamma_i}]_{i \in \{0, \ldots, n/8-1\}}} \left[\prod \max_{\substack{\beta \in \{0,1\} \ \mathbf{x}_i \sim \mathcal{S}_{\gamma_i} \\ b \sim \{0,1\}}} \Pr \left[\mathsf{Rec}(\mathbf{x}_i, \mathbf{r}_i) = \beta \mid \mathsf{HelpRec}(\mathbf{x}_i; b) = \mathbf{r}_i \right] \right]
$$

$$
= \sum_{\mathbf{y}} E_{[\mathbf{r}_i \sim \mathcal{R}_{\gamma_i}]_{i \in \{0, \ldots, n/8-1\}}} \left[\prod \max_{\substack{\beta \in \{0,1\} \ \mathbf{x}_i \sim \mathcal{S}_{\gamma_i} \\ b \sim \{0,1\}}} \Pr \left[\mathsf{Rec}(\mathbf{x}_i, \mathbf{r}_i) = \beta \mid \mathsf{HelpRec}(\mathbf{x}_i; b) = \mathbf{r}_i \right] \right] \cdot \Pr_{\Psi}[\mathbf{y}]
$$

$$
\leq \sum_{\mathbf{y}} E_{[\mathbf{r}_i \sim \mathcal{R}_{\gamma_i}]_{i \in \{0, \ldots, n/8-1\}}} \left[\prod \max_{\substack{\beta \in \{0,1\} \ \mathbf{x}_i \sim \mathcal{S}_{\gamma_i} \\ b \sim \{0,1\}}} \Pr \left[\mathsf{Rec}(\mathbf{x}_i, \mathbf{r}_i) = \beta \mid \mathsf{HelpRec}(\mathbf{x}_i; b) = \mathbf{r}_i \right] \right]
$$

$$
\cdot \frac{(1 + 1/166)^{128}}{q^{n/8}}
$$

$$
= (1 + 1/166)^{128} E_{\mathbf{y} \leftarrow \mathbb{Z}_q^{n/8}, [\mathbf{r}_i \sim \mathcal{R}_{\gamma_i}]_{i \in \{0, \ldots, n/8-1\}}}
$$

$$
\left[\prod \max_{\substack{\beta \in \{0,1\} \ \mathbf{x}_i \sim \mathcal{S}_{\gamma_i} \\ b \sim \{0,1\}}} \Pr \left[\mathsf{Rec}(\mathbf{x}_i, \mathbf{r}_i) = \beta \mid \mathsf{HelpRec}(\mathbf{x}_i; b) = \mathbf{r}_i \right] \right]
$$

$$= (1 + 1/166)^{128} \prod_i E_{\gamma_i \leftarrow Z_q^{n/8}, [r_i \sim \mathcal{R}_{\gamma_i}]_{i \in \{0, \ldots, n/8-1\}}}$$

$$\left[\max_{\substack{\beta \in \{0,1\} \ x_i \sim \mathcal{S}_{\gamma_i} \\ b \sim \{0,1\}}} \Pr [\mathsf{Rec}(x_i, r_i) = \beta \mid \mathsf{HelpRec}(x_i; b) = r] \right].$$

Thus, we can lower bound the average min-entropy of the output of Rec for all blocks $i \in \{0, \ldots, n/8 - 1\}$ by analyzing the expectation for a single block $E_{\gamma_i \leftarrow Z_q^{n/8}, [r_i \sim \mathcal{R}_{\gamma_i}]_{i \in \{0, \ldots, n/8-1\}}} \left[\max_{\beta \in \{0,1\}} \Pr_{\substack{x_i \sim \mathcal{S}_{\gamma_i} \\ b \sim \{0,1\}}} [\mathsf{Rec}(x_i, r_i)$

$= \beta \mid \mathsf{HelpRec}(x_i; b) = r_i] \Big]$, taking the negative log, multiplying by 128 (the

number of blocks) and subtracting $\log_2(1 + 1/166)^{128} \approx 1.1091$.

Remark 4.2.1 In the following, we drop the subscript i from the variables x_i, r_i, γ_i, since we focus on a single block at a time.

Theorem 4.9 *We have that:*

$$E_{\gamma \leftarrow Z_q, r \sim \mathcal{D}_\gamma} \left[\max_{\substack{\beta \in \{0,1\} \ x \sim \mathcal{S}_\gamma \\ b \sim \{0,1\}}} \Pr [\mathsf{Rec}(x, r) = \beta \mid \mathsf{HelpRec}(x; b) = r] \right] \leq 1/2 + p/2,$$

where

$$p := 2 - 2 \left(\frac{\frac{q}{2^r} - 2q^{1/4} - 1}{\frac{q}{2^r}} \right)^4 + \left(\frac{1 + \frac{1}{2q^{1/4}}}{1 - \frac{2^{r+1}}{q}} \right)^4 \cdot \left(\frac{2^{r+10}}{3q^{3/4}} - \frac{2^{3r+10}}{3q^{9/4}} + \frac{2^{4r+10}}{q^3} \right).$$

Proof We prove the theorem by showing that for linear constraint $c_{\omega, \alpha}$, there exists a bijective mapping $\psi_{c_{\omega, \alpha}}(x) = x'$, such that, with high probability at least $1 - p$ over uniform x, all the following conditions hold:

$$c_{\omega, \alpha} \cdot x = c_{\omega, \alpha} \cdot x' \tag{3}$$

$$(r =) \mathsf{HelpRec}(x; b) = \mathsf{HelpRec}(x'; b'), \tag{4}$$

$$\mathsf{Rec}(x, r) = 1 \oplus \mathsf{Rec}(x', r), \tag{5}$$

where $b' = b \oplus 1$. $\qquad \square$

Now the above conditions imply that:

$$\frac{1}{q} \cdot \sum_{(\gamma, \mathbf{r})} \Pr_{\mathcal{R}_\gamma}[\mathbf{r}] \cdot \Pr_{\substack{\mathbf{x} \sim \mathcal{S}_\gamma \\ b \sim \{0,1\}}} [\mathsf{HelpRec}(\mathbf{x}; b) \neq \mathsf{HelpRec}(\mathbf{x}'; b') \mid \mathsf{HelpRec}(\mathbf{x}; b) = \mathbf{r}] \leq p.$$

(6)

Let $p_{\gamma, \mathbf{r}} := \Pr_{\substack{\mathbf{x} \sim \mathcal{S}_\gamma \\ b \sim \{0,1\}}} [\mathsf{HelpRec}(\mathbf{x}; b) \neq \mathsf{HelpRec}(\mathbf{x}'; b') \mid \mathsf{HelpRec}(\mathbf{x}; b) =$
$\mathbf{r}]$. Note that $\max_{\beta \in \{0,1\}} \Pr_{\substack{\mathbf{x} \sim \mathcal{S}_\gamma \\ b \sim \{0,1\}}} [\mathsf{Rec}(\mathbf{x}, \mathbf{r}) = \beta \mid \mathsf{HelpRec}(\mathbf{x}; b) = \mathbf{r}] \leq 1/2 +$
$p_{\gamma, \mathbf{r}}/2$.

This is sufficient to prove Theorem 4.9, since

$$E_{\gamma \leftarrow Z_q, \mathbf{r} \sim \mathcal{D}_\gamma}[\max_{\beta \in \{0,1\}} \Pr_{b \sim \{0,1\}, \mathbf{x} \sim \mathcal{S}_\gamma} [\mathsf{Rec}(\mathbf{x}, \mathbf{r}) = \beta \mid \mathsf{HelpRec}(\mathbf{x}; b) = \mathbf{r}]]$$

$$\leq \frac{1}{q} \cdot \sum_{(\gamma, \mathbf{r})} \Pr_{\mathcal{R}_\gamma}[\mathbf{r}] \cdot (1/2 + p_{\gamma, \mathbf{r}}/2)$$

$$= 1/2 + \frac{1}{q} \cdot \sum_{(\gamma, \mathbf{r})} \Pr_{\mathcal{R}_\gamma}[\mathbf{r}] \cdot p_{\gamma, \mathbf{r}}/2$$

$$= \frac{1}{2} + \frac{1}{2q} \sum_{(\gamma, \mathbf{r})} \Pr_{\mathcal{R}_\gamma}[\mathbf{r}] \cdot \Pr_{\substack{\mathbf{x} \sim \mathcal{S}_\gamma \\ b \sim \{0,1\}}} [\mathsf{HelpRec}(\mathbf{x}; b) \neq \mathsf{HelpRec}(\mathbf{x}'; b')$$

$$\mid \mathsf{HelpRec}(\mathbf{x}; b) = \mathbf{r}]$$

$$\leq 1/2 + p/2,$$

where the last inequality follows from (6).

We now turn to defining $\psi_{c_{\omega, \alpha}}$ and proving that with probability at least $1 - p$ over uniform $\mathbf{x}$, (3), (4), and (5) hold.

Defining $\psi_{c_{\omega, \alpha}}$ so that (3) Always Holds Equation (3) holds if and only if there exists a vector $\mathbf{w} \in \mathbb{Z}_q^4$ such that $\mathbf{x}' = \mathbf{x} + \mathbf{w}$, where $\mathbf{w} \in \ker(c_{\omega, \alpha})$, where ker is the set of $\mathbf{w}'$ such that $c_{\omega, \alpha} \cdot \mathbf{w}' = 0$. Let $\mathbf{W}$ to be a set of all vectors $\mathbf{vt} = (vt_0, vt_1, vt_2, vt_3)$ where $vt_i \in [\frac{q}{2} \pm q^{1/4}] \cap \mathbb{Z}$. By conducting an exhaustive search, we observe that the intersection of set $\ker(c_{\omega, \alpha})$ and set $\mathbf{W}$ is nonempty given parameter setting of [3], namely fixing $q = 12289, n = 1024, \omega = 7$, $\ker(f_{\omega, j}) \cap \mathbf{W} \neq \emptyset$ for all $\alpha \in \mathbb{Z}_{16}^*$.[7] Define $\psi_{c_{\omega, \alpha}}(\mathbf{x}) := \mathbf{x} + \mathbf{w}$, where $\mathbf{w} \in \ker(c_{\omega, \alpha}) \cap \mathbf{W}$. Therefore, as long as $\ker(c_{\omega, \alpha}) \cap \mathbf{W}$ is non-empty (which holds for typical parameter settings), condition (3) holds with probability 1 over choice of $\mathbf{x}$.

---

[7]Note that this is the only part of the analysis that is not generic in terms of parameter settings. For more discussion, See Sect. 4.4.

If (4) Holds then (5) Holds We now show that if $\mathbf{x}$ is such that $\mathsf{HelpRec}(\mathbf{x}; b) = \mathsf{HelpRec}(\mathbf{x} + \mathbf{w}; b')$ then if $\mathsf{HelpRec}(\mathbf{x}; b) = \mathbf{r}$, $\mathsf{Rec}(\mathbf{x}, \mathbf{r}) = 1 \oplus \mathsf{Rec}(\mathbf{x} + \mathbf{w}, \mathbf{r})$.

Lemma 4.10 *Given* $\mathsf{HelpRec}(\mathbf{x}; b) = \mathsf{HelpRec}(\mathbf{x} + \mathbf{w}; b') = \mathbf{r}$, *we have* $\mathsf{Rec}(\mathbf{x}, \mathbf{r}) = 1 \oplus \mathsf{Rec}(\mathbf{x} + \mathbf{w}, \mathbf{r})$.

Proof Recall that $\mathbf{g} = (1/2, 1/2, 1/2, 1/2)^T$. Proved by [5, Lemma C.2], we have

$$\mathsf{HelpRec}(\mathbf{x}; b) = \mathsf{HelpRec}(\mathbf{x} + q\mathbf{g}) \ (= \mathbf{r})$$

$$\mathsf{Rec}(\mathbf{x}, \mathbf{r}) = 1 \oplus \mathsf{Rec}(\mathbf{x} + q\mathbf{g}, \mathbf{r})$$

Additionally, since $\|\mathbf{w} - q\mathbf{g}\|_1 \leq 4q^{1/4} < (1 - 1/2^r) \cdot q - 2$, by [5, Lemma C.3], $\mathsf{Rec}(\mathbf{x}+\mathbf{w}, \mathbf{r}) = \mathsf{Rec}(\mathbf{x}+q\mathbf{g}, \mathbf{r})$, Thus we conclude $\mathsf{Rec}(\mathbf{x}, \mathbf{r}) = 1\oplus\mathsf{Rec}(\mathbf{x}+\mathbf{w}, \mathbf{r})$. $\qquad\square$

Equation (4) Holds with Probability $1 - p$ ***Over*** $\mathbf{x}$ Hence, it remains to show that for all $\mathbf{w} \in \ker(c_{\omega,\alpha}) \cap \mathbf{W}$ and $f_{\omega,j}(\mathbf{x}) = \gamma$, with high probability at least $1 - p$ over choice of $\mathbf{x} \leftarrow \mathbb{Z}_q^4$, $b \leftarrow \{0, 1\}$, $\mathsf{HelpRec}(\mathbf{x}; b) = \mathsf{HelpRec}(\mathbf{x} + \mathbf{w}; b')$ holds.

Let $\boldsymbol{\delta} = (\delta_0, \delta_1, \delta_2, \delta_3)$ be a vector such that $\mathbf{x} + \mathbf{w} = \mathbf{x} + q\mathbf{g} + \boldsymbol{\delta}$. Then $|\delta_i| \leq q^{1/4}$. Since $\mathbf{g} \in \tilde{D}_4$, we have $\mathsf{HelpRec}(\mathbf{x}; b) = \mathsf{HelpRec}(\mathbf{x} + q\mathbf{g}; b')$ [3]. For simplicity, let $\mathbf{z} = \frac{2^r}{q}(\mathbf{x} + q\mathbf{g} + b'\mathbf{g}) \in \frac{2^r}{2q}\mathbb{Z}_{2q}^4$, vector $\boldsymbol{\beta} = (\beta_0, \beta_1, \beta_2, \beta_3)$ denote $\frac{2^r}{q}\boldsymbol{\delta}$. Recall that $\mathsf{HelpRec}(\mathbf{x}; b) = \mathsf{CVP}_{\tilde{D}_4}\left(\frac{2^r}{q}(\mathbf{x} + b\mathbf{g})\right) \bmod 2^r$. Thus, the proposition $(\mathsf{HelpRec}(\mathbf{x} + q\mathbf{g}; b') =)\mathsf{HelpRec}(\mathbf{x}; b) = \mathsf{HelpRec}(\mathbf{x} + \mathbf{w}; b')$ is equivalent to $\mathsf{CVP}_{\tilde{D}_4}(\mathbf{z}) = \mathsf{CVP}_{\tilde{D}_4}(\mathbf{z} + \boldsymbol{\beta})$, which remains to be proved valid with probability at least $1 - p$.

For the following analysis, refer to Algorithm 4.3, which describes the $\mathsf{CVP}_{\tilde{D}_4}$ algorithm. Let $\boldsymbol{v}_0, \boldsymbol{v}_1, k$ be the values computed in steps 1, 2, 3 of $\mathsf{CVP}_{\tilde{D}_4}(\mathbf{z})$ algorithm, shown in Algorithm 4.3 and let $\boldsymbol{v}_0', \boldsymbol{v}_1', k'$ be the values computed in step 1, 2, 3 of $\mathsf{CVP}_{\tilde{D}_4}(\mathbf{z} + \boldsymbol{\beta})$ algorithm.

By definition of $\mathsf{CVP}_{\tilde{D}_4}$, it is clear to see that if none of the following three conditions is satisfied, then $\mathsf{CVP}_{\tilde{D}_4}(\mathbf{z}) = \mathsf{CVP}_{\tilde{D}_4}(\mathbf{z} + \boldsymbol{\beta})$ is granted.

(a) $\boldsymbol{v}_0' \neq \boldsymbol{v}_0$.
(b) $\boldsymbol{v}_1' \neq \boldsymbol{v}_1$
(c) $k' \neq k$

Before analyzing probability in each condition above, we first present the following lemma, which will allow us to switch from analyzing the probabilities over choice of $(\mathbf{x}, b)$ to analyzing probabilities over choice of $\mathbf{z}$.

Lemma 4.11 *Given* $\mathbf{g}, \mathbf{x}, \mathbf{z} = \frac{2^r}{q}(\mathbf{x} + q\mathbf{g} + b'\mathbf{g})$ *as defined above, for any set* $\mathcal{D}' \subseteq \frac{2^r}{2q}\mathbb{Z}_{2q}^4$, *the probability that* $\mathbf{x}$ *in set* $\mathcal{D} = \{\mathbf{x} \mid \frac{2^r}{q}(\mathbf{x} + q\mathbf{g} + b'\mathbf{g}) \in \mathcal{D}'\}$ *over choice of* $\mathbf{x} \leftarrow \mathbb{Z}_q^4$ *and choice of* $b' \leftarrow \{0, 1\}$ *equals to the probability that* $\mathbf{z}$ *in set* $\mathcal{D}'$ *over choice of* $\mathbf{z} \leftarrow \frac{2^r}{2q}\mathbb{Z}_{2q}^4$, *denoted by* $\mathrm{Prob}_{\mathbf{x},b'}[\mathbf{x} \in \mathcal{D}] = \mathrm{Prob}_{\mathbf{z}}[\mathbf{z} \in \mathcal{D}']$.

Proof We compute $\mathrm{Prob}_{\mathbf{x},b'}[\mathbf{x} \in \mathcal{D}]$ given the condition $b' = 0$ and the condition $b' = 1$. As b' is equivalent to the "doubling" trick (See [44] for example), the corresponding $\mathbf{x} + q\mathbf{g} + b'\mathbf{g}$ when $b' = 0$ is distributed as odd numbers over $\mathbb{Z}_{2q}^4$, written as $2\mathbb{Z}_{2q}^4 + \mathbb{Z}_{2q}^4$. When $b' = 1$, $\mathbf{x} + q\mathbf{g} + b'\mathbf{g}$ is distributed as even numbers over is over $\mathbb{Z}_{2q}^4$, written as $2\mathbb{Z}_{2q}^4$. Thus we have

$$\mathrm{Prob}_{\mathbf{x},b'}[\mathbf{x} \in \mathcal{D}] = \frac{1}{2}\mathrm{Prob}_{\mathbf{x},0}\left[\frac{2^r}{q}(\mathbf{x} + q\mathbf{g}) \in \mathcal{D}'\right] + \frac{1}{2}\mathrm{Prob}_{\mathbf{x},1}\left[\frac{2^r}{q}(\mathbf{x} + q\mathbf{g} + \mathbf{g}) \in \mathcal{D}'\right]$$

(7)

$$= \frac{1}{2}\frac{\left|\frac{2^r}{2q}(2\mathbb{Z}_{2q}^4 + \mathbb{Z}_{2q}^4) \cap \mathcal{D}'\right|}{\left|\frac{2^r}{2q}(2\mathbb{Z}_{2q}^4 + \mathbb{Z}_{2q}^4)\right|} + \frac{1}{2}\frac{\left|\frac{2^r}{2q}(2\mathbb{Z}_{2q}^4) \cap \mathcal{D}'\right|}{\left|\frac{2^r}{2q}(2\mathbb{Z}_{2q}^4)\right|}$$

(8)

$$= \frac{\left|\left(\frac{2^r}{2q}(2\mathbb{Z}_{2q}^4 + \mathbb{Z}_{2q}^4) \cup \frac{2^r}{2q}(2\mathbb{Z}_{2q}^4)\right) \cap \mathcal{D}'\right|}{\frac{2^r}{2q}(\mathbb{Z}_{2q}^4)}$$

(9)

$$= \frac{\left|\frac{2^r}{2q}(\mathbb{Z}_{2q}^4) \cap \mathcal{D}'\right|}{\frac{2^r}{2q}(\mathbb{Z}_{2q}^4)} = \mathrm{Prob}_{\mathbf{z}}[\mathbf{z} \in \mathcal{D}']$$

(10)

as desired.   □

We omit to mention distribution of b' for simplicity.

We next analyze probability of the three conditions (a), (b), and (c) in Lemmas 4.12, 4.13, and 4.14.

Lemma 4.12 (Bounding the Probability of (a)) *Given v_0, v_0', v_1, v_1', k, k', $\mathbf{z}$, β as defined above, probability that $v_0' \neq v_0$ (denoted by $\mathrm{Prob}_{\mathbf{x}}[v_0' \neq v_0]$) is at most*

$$1 - \left(\frac{\frac{q}{2^r} - 2q^{1/4} - 1}{\frac{q}{2^r}}\right)^4 \text{ over choice of } \mathbf{x} \leftarrow \mathbb{Z}_q^4.$$

Proof Recall that $|\delta_i| \leq q^{1/4}$. Then we have $|\beta_i| \leq \frac{2^r}{q^{3/4}}$. We assume that $\frac{2^r}{q^{3/4}} \leq 1/2$, which would be the case for typical parameter settings (for example $r = 2$, $q = 12289$). When the event that $v_0' \neq v_0$ happens, it indicates existing an i such that, $\lfloor z_i \rceil \neq \lfloor z_i + \beta_i \rceil$. We start by computing the probability over choice of $\mathbf{x} \leftarrow \mathbb{Z}_q^4$ that given i, event $\lfloor z_i \rceil = \lfloor z_i + \beta_i \rceil$ occurs, denoted by $\mathrm{Prob}_{\mathbf{x}}[\lfloor z_i \rceil = \lfloor z_i + \beta_i \rceil]$. We divide the analysis into two cases.

(1) Suppose that $z_i - \lfloor z_i \rceil \geq 0$, then $\lfloor z_i - \frac{2^r}{q^{3/4}} \rceil = \lfloor z_i \rceil$. In order to achieve $\lfloor z_i + \beta_i \rceil = \lfloor z_i \rceil$, we need $\lfloor z_i + \frac{2^r}{q^{3/4}} \rceil = \lfloor z_i \rceil$. Without loss of generality, we assume $0 \leq z_i < 1/2 \bmod 2^r$, where $\lfloor z_i \rceil = 0$. Thus it can be easily verified that if $0 \leq z_i < 1/2 - \frac{2^r}{q^{3/4}}$, we can ensure $\lfloor z_i + \frac{2^r}{q^{3/4}} \rceil = 0$.

(2) Suppose that $z_i - \lceil z_i \rceil < 0$, then $\lceil z_i + \frac{2^r}{q^{3/4}} \rceil = \lceil z_i \rceil$. Similarly, in order to achieve $\lceil z_i + \beta_i \rceil = \lceil z_i \rceil$, we need $\lceil z_i - \frac{2^r}{q^{3/4}} \rceil = \lceil z_i \rceil$. Without loss of generality, we assume $-1/2 \leq z_i < 0 \mod 2^r$, where $\lceil z_i \rceil = 0$. Thus it can easily verified that if $-1/2 + \frac{2^r}{q^{3/4}} \leq z_i < 0$, we can ensure $\lceil z_i + \frac{2^r}{q^{3/4}} \rceil = 0$.

Combining both cases, by Lemma 4.11, we then derive that

$$\mathrm{Prob_x}[\lceil z_i \rceil = \lceil z_i + \beta_i \rceil] \geq \frac{\left| \left[-1/2 + \frac{2^r}{q^{3/4}}, 1/2 - \frac{2^r}{q^{3/4}} \right) \cap \frac{2^r}{2q} \mathbb{Z}_{2q} \right|}{\left| [-1/2, 1/2) \cap \frac{2^r}{2q} \mathbb{Z}_{2q} \right|} \tag{11}$$

$$\geq \frac{\left| \left[\frac{2q}{2^r}(-1/2 + \frac{2^r}{q^{3/4}}), \frac{2q}{2^r}(1/2 - \frac{2^r}{q^{3/4}}) \right) \cap \mathbb{Z}_{2q} \right|}{\left| \left[-\frac{q}{2^r}, \frac{q}{2^r} \right) \cap \mathbb{Z}_{2q} \right|} \tag{12}$$

$$= \frac{2 \lfloor \frac{q}{2^r} - 2q^{1/4} \rfloor + 1}{2 \lfloor \frac{q}{2^r} \rfloor + 1} \tag{13}$$

$$\geq \frac{\frac{q}{2^r} - 2q^{1/4} - 1}{\frac{q}{2^r}} \tag{14}$$

Since $\mathrm{Prob_x}[\exists i, \lfloor z_i \rceil \neq \lfloor z_i + \beta_i \rceil] = 1 - \mathrm{Prob_x}[\lfloor z_i \rceil = \lfloor z_i + \beta_i \rceil, \forall i]$. Therefore, we have

$$\mathrm{Prob_x}[v'_0 \neq v_0] \leq 1 - \left(\frac{\frac{q}{2^r} - 2q^{1/4} - 1}{\frac{q}{2^r}} \right)^4$$

as desired. □

Lemma 4.13 (Bounding the Probability of (b)) *Given $v_0, v'_0, v_1, v'_1, k, k', \mathbf{z}, \beta$ as defined above, probability that $v'_1 \neq v_1$ (denoted by $\mathrm{Prob_x}[v'_1 \neq v_1]$) is at most $1 - \left(\frac{\frac{q}{2^r} - 2q^{1/4} - 1}{\frac{q}{2^r}} \right)^4$ over choice of $\mathbf{x} \leftarrow \mathbb{Z}_q^4$.*

The proof proceeds exactly the same as proof of Lemma 4.12 by substituting $\mathbf{z}$ with $\mathbf{z} + \mathbf{g}$.

Lemma 4.14 (Bounding the Probability of (c)) *Given $v_0, v'_0, v_1, v'_1, k, k', \mathbf{z}, \beta$ as defined above, probability that $k' \neq k$ (denoted by $\mathrm{Prob_x}[k' \neq k]$) is at most $\left(\frac{1 + \frac{1}{2q^{1/4}}}{1 - \frac{2^{r+1}}{q}} \right)^4 \cdot \left(\frac{2^{r+10}}{3q^{3/4}} - \frac{2^{3r+10}}{3q^{9/4}} + \frac{2^{4r+10}}{q^3} \right)$ over choice of $\mathbf{x} \leftarrow \mathbb{Z}_q^4$.*

Proof We divide our proof into two cases: (1) Suppose $k = 0$ and $k' = 1$, which indicates $\|\mathbf{z} - v_0\|_1 < 1$ and $\|\mathbf{z} + \beta - v'_0\|_1 \geq 1$. We denote by $\mathrm{Prob_x}[k = 0, k' = 1]$ probability that $k = 0$ and $k' = 1$ over choice of $\mathbf{x}$. (2) Suppose $k = 1$ and

$k' = 0$, which indicates $\|\mathbf{z} - \boldsymbol{v}_0\|_1 \geq 1$ and $\|\mathbf{z} + \boldsymbol{\beta} - \boldsymbol{v}_0'\|_1 < 1$. We denote by $\mathrm{Prob}_{\mathbf{x}}[k = 1, k' = 0]$ probability that $k = 1$ and $k' = 0$ over choice of $\mathbf{x}$.

Without loss of generality, we assume that $-1/2 \leq z_i < 1/2 \mod 2^r$ for $i = 0, 1, 2, 3$. Then we have $\boldsymbol{v}_0 = \mathbf{0}$.

Case 1 By Lemma 4.11, $\mathrm{Prob}_{\mathbf{x}}[k = 0, k' = 1]$ is equivalent to the probability that $\mathbf{z}$ satisfies

$$|z_0| + |z_1| + |z_2| + |z_3| < 1, \text{ and}$$

$$|z_0 + \beta_0| + |z_1 + \beta_1| + |z_2 + \beta_2| + |z_3 + \beta_3| > 1,$$

over choice of $\mathbf{z} \leftarrow \frac{2^r}{2q}\mathbb{Z}_{2q}^4 \cap [-1/2, 1/2)^4 \mod 2^r$. As $|z_i + \beta_i| \leq |z_i| + |\beta_i|$ by Triangle Inequality, we can upper-bound $\mathrm{Prob}_{\mathbf{x}}[k = 0, k' = 1]$ by the probability that $\mathbf{z}$ satisfies

$$|z_0| + |z_1| + |z_2| + |z_3| < 1, \text{ and}$$

$$|z_0| + |z_1| + |z_2| + |z_3| + |\beta_0| + |\beta_1| + |\beta_2| + |\beta_3| > 1,$$

over choice of $\mathbf{z} \leftarrow \frac{2^r}{2q}\mathbb{Z}_{2q}^4 \cap [-1/2, 1/2)^4 \mod 2^r$. Since $|\beta_0| + |\beta_1| + |\beta_2| + |\beta_3| \leq 4 \cdot \frac{2^r}{q^{3/4}}$, we can further upper-bound $\mathrm{Prob}_{\mathbf{x}}[k = 0, k' = 1]$ by the probability that $\mathbf{z}$ satisfies

$$1 - 4 \cdot \frac{2^r}{q^{3/4}} < |z_0| + |z_1| + |z_2| + |z_3| < 1,$$

over choice of $\mathbf{z} \leftarrow \frac{2^r}{2q}\mathbb{Z}_{2q}^4 \cap [-1/2, 1/2)^4 \mod 2^r$.

Let $\Delta = 4 \cdot \frac{2^r}{q^{3/4}}$. We then can obtain the upperbound of $\mathrm{Prob}_{\mathbf{x}}[k = 0, k' = 1]$ by computing the cardinality of set where each element is in $\frac{2^r}{2q}\mathbb{Z}_{2q}^4$ and satisfies the following two conditions:

$$1 - \Delta < |z_0| + |z_1| + |z_2| + |z_3| < 1 \tag{15}$$

$$-1/2 \leq z_i < 1/2 \text{ , for } i = 0, 1, 2, 3 \tag{16}$$

divided by the cardinality of set where each element is in $\frac{2^r}{2q}\mathbb{Z}_{2q}^4$ and only satisfies the Eq. (16).

Similarly for *Case 2*, by Lemma 4.11, $\mathrm{Prob}_{\mathbf{x}}[k = 1, k' = 0]$ is equivalent to the probability that $\mathbf{z}$ satisfies

$$|z_0| + |z_1| + |z_2| + |z_3| \geq 1, \text{ and}$$

$$|z_0 + \beta_0| + |z_1 + \beta_1| + |z_2 + \beta_2| + |z_3 + \beta_3| < 1,$$

over choice of $\mathbf{z} \leftarrow \frac{2^r}{2q}\mathbb{Z}_{2q}^4 \cap [-1/2, 1/2)^4 \mod 2^r$. Since $|z_i + \beta_i| \geq |z_i| - |\beta_i|$ and $|\beta_0| + |\beta_1| + |\beta_2| + |\beta_3| \leq 4 \cdot \frac{2^r}{q^{3/4}}$, we can further upper -bounded $\mathrm{Prob}_\mathbf{x}[k = 1, k' = 0]$ by the probability that $\mathbf{z}$ satisfies

$$1 \leq |z_0| + |z_1| + |z_2| + |z_3| < 1 + 4 \cdot \frac{2^r}{q^{3/4}},$$

over choice of $\mathbf{z} \leftarrow \frac{2^r}{q}\mathbb{Z}_q^4 \cap [-1/2, 1/2)^4 \mod 2^r$.

We can then obtain the upperbound of $\mathrm{Prob}_\mathbf{x}[k = 1, k' = 0]$ by computing the cardinality of set where each element is in $\frac{2^r}{2q}\mathbb{Z}_{2q}^4$ and satisfies the following two conditions:

$$1 \leq |z_0| + |z_1| + |z_2| + |z_3| < 1 + \Delta \tag{17}$$

$$-1/2 \leq z_i < 1/2, \text{for} i = 0, 1, 2, 3 \tag{18}$$

by the cardinality of set that each element is in $\frac{2^r}{2q}\mathbb{Z}_{2q}^4$ and only satisfies Eq. (18).

Thus, by combining both cases, we have $\mathrm{Prob}_\mathbf{x}[k' \neq k] = \mathrm{Prob}_\mathbf{x}[k = 0, k' = 1] + \mathrm{Prob}_\mathbf{x}[k = 1, k' = 0]$ upperbounded by the cardinality of set where each element is in $\frac{2^r}{2q}\mathbb{Z}_{2q}^4$ and satisfies the following two conditions:

$$1 - \Delta < |z_0| + |z_1| + |z_2| + |z_3| < 1 + \Delta \tag{19}$$

$$-1/2 \leq z_i < 1/2, \text{for} i = 0, 1, 2, 3 \tag{20}$$

by the cardinality of set where elements are in $\frac{2^r}{2q}\mathbb{Z}_{2q}^4$ and satisfies the Eq. (20).

Note that, disregarding the distribution of $\mathbf{z}$, (20) defines a unit hypercube $[-1/2, 1/2)^4$ centered at origin and (19) defines a hyper-object clipped by two hyperplanes in each octant. We denote by Vol_{cube} the hypercube volume. Let Vol_{clip} be the hypervolume where each points satisfies both (19) and (20), which is equivalent to say, Vol_{clip} is hypervolume of hypercube defined in (20) clipped by two hyperplanes in each octant, as defined in (19).

If $\mathbf{x}$ is sampled from $\mathbb{R}^4$, it is easy to see that probability $\mathrm{Prob}_\mathbf{x}[k' \neq k]$ is upperbounded by the ratio of Vol_{clip} to Vol_{cube}.

For the rest of the proof, we first compute the ratio of Vol_{clip} to Vol_{cube} and then approximate the upperbound of $\mathrm{Prob}_\mathbf{x}[k' \neq k]$ by discretizing hypervolumes into lattice points, as $\mathbf{z}$ is instead sampled from $\frac{2^r}{2q}\mathbb{Z}_{2q}^4$, which is a lattice.

Towards computing the volumes, we need to amplify each dimension by 2 in (15) and (16) for adapting Theorem 2.2 where unit hypercube is defined as $[0, 1]^n$. Vol_{clip} and Vol_{cube} is ith octant. Let Vol_{clip}^i denote Vol_{clip} in the ith octant, and Vol_{cube}^i denote Vol_{cube} in ith octant. Thus, in the ith octant, we have

$$2 - 2\Delta < t_0 + t_1 + t_2 + t_3 < 2 + 2\Delta \tag{21}$$

$$0 \leq t_i < 1, \text{for} i = 0, 1, 2, 3, \tag{22}$$

where $t_i = 2z_i$.

Define two hyperspace as follows:

$$H_1^+ := \{\mathbf{t} \mid g_1(\mathbf{t}) := -t_0 - t_1 - t_2 - t_3 + 2(1 - \Delta) \geq 0\}$$

$$H_2^+ := \{\mathbf{t} \mid g_2(\mathbf{t}) := -t_0 - t_1 - t_2 - t_3 + 2(1 + \Delta) \geq 0\}$$

Then it is easy to see that

$$\text{Vol}_{clip}^1 \leq \frac{\text{Vol}([0, 1]^4 \cap H_2^+) - \text{Vol}([0, 1]^4 \cap H_1^+)}{2^4}.$$

where 2^4 in denominator is a scalar to neutralize amplification

By Theorem 2.2 and substituting Δ with $4q^{1/4} \cdot \frac{2^r}{q}$, we obtain

$$\text{Vol}_{clip}^1 = \frac{1}{2^4} \cdot \frac{1}{24} \left((2 + 2\Delta)^4 - 4(1 + 2\Delta)^4 + 6(2\Delta)^4 - (2 - 2\Delta)^4 + 4(1 - 2\Delta)^4 \right)$$

$$= \frac{1}{2^4} \cdot \frac{1}{24} (64\Delta - 128\Delta^3 + 96\Delta^4)$$

$$= \frac{1}{2^4} \left(\frac{2^{4r+10}}{q^3} - \frac{2^{3r+10}}{3q^{9/4}} + \frac{2^{r+5}}{3q^{3/4}} \right)$$

We claim that $\text{Vol}_{clip}^1 = \text{Vol}_{clip}^i$ for $i = 2, 3, \ldots, 16$. It can be easily checked by showing a bijective map $f_i : \mathbf{z} \leftrightarrow \mathbf{z}'$ which maps elements from first octant to the ith octant, such that if $\mathbf{z}$ satisfies the conditions (19) and (20), then $\mathbf{z}'$ satisfies the conditions (19) and (20), and if $\mathbf{z}$ satisfies the condition (20) but not satisfies (19), then $\mathbf{z}'$ satisfies the condition (20) but not satisfies (19). One trivial example of such map is to let $\mathbf{z}$ be the absolute value of $\mathbf{z}'$.

Additionally, it is obvious to see that $\text{Vol}_{cube}^i = 1/2^4, \forall i$. Thus, we have

$$\frac{\text{Vol}_{clip}}{\text{Vol}_{cube}} = \frac{\text{Vol}_{clip}^1}{\text{Vol}_{cube}^1} = \frac{2^{4r+10}}{q^3} - \frac{2^{3r+10}}{3q^{9/4}} + \frac{2^{r+5}}{3q^{3/4}}.$$

It remains to approximate $\frac{\text{Vol}_{clip}^1 \cap \mathcal{L}_{\mathbf{z}}}{\text{Vol}_{cube}^1 \cap \mathcal{L}_{\mathbf{z}}}$, where $\mathcal{L}_{\mathbf{z}} = \frac{2^r}{2q}\mathbb{Z}_{2q}^4$.

Since both of the hypercube and the hyperclip in first octant are convex as they are intersections of hyperspaces, by Theorem 2.1, we can derive that

$$\frac{\text{Vol}_{clip}^1 \cap \mathcal{L}_{\mathbf{z}}}{\text{Vol}_{cube}^1 \cap \mathcal{L}_{\mathbf{z}}} \leq \frac{(1 + \varepsilon)^4 \frac{\text{Vol}_{clip}^1}{\det(\mathcal{L}_{\mathbf{z}})}}{(1 - \varepsilon')^4 \frac{\text{Vol}_{cube}^1}{\det(\mathcal{L}_{\mathbf{z}})}} = \left(\frac{1 + \varepsilon}{1 - \varepsilon'} \right)^4 \cdot \frac{\text{Vol}_{clip}^1}{\text{Vol}_{cube}^1},$$

where $\mathcal{P}(B) \cup -\mathcal{P}(B) \subseteq \varepsilon \cdot \text{Vol}^1_{clip}$, $\mathcal{P}(B) \cup -\mathcal{P}(B) \subseteq \varepsilon' \cdot \text{Vol}^1_{cube}$ and B is a basis of $\mathcal{L}_{\mathbf{z}}$.

To get a small ε, we begin by carving a hypercube $[\frac{1}{4} - \frac{1}{4}\Delta, \frac{1}{4} + \frac{1}{4}\Delta]^4$, which is contained in Vol^1_{clip}. Let $B = \{(\frac{2^r}{2q}, 0, 0, 0), (0, \frac{2^r}{2q}, 0, 0), (0, 0, \frac{2^r}{2q}, 0), (0, 0, 0, \frac{2^r}{2q})\}$. Then $\mathcal{P}(B)$ forms a hypercube with side length $\frac{2^r}{2q}$. Thus, by letting $\varepsilon = \frac{1}{2q^{1/4}}$ as $\frac{2^r}{2q} \cdot 2 \leq \varepsilon \cdot \frac{1}{2}\Delta$, we can guarantee that $\mathcal{P}(B) \cup -\mathcal{P}(B) \subseteq \varepsilon \cdot \text{Vol}^1_{clip}$. Similarly, since Vol^1_{cube} is a hypercube, it is easy to see that by letting $\varepsilon' = \frac{2^{r+1}}{q}$, $\mathcal{P}(B) \cup -\mathcal{P}(B) \subseteq \varepsilon' \cdot \text{Vol}^1_{cube}$.

Combining the above, we obtain

$$\frac{\text{Vol}^1_{clip} \cap \mathcal{L}_{\mathbf{z}}}{\text{Vol}^1_{cube} \cap \mathcal{L}_{\mathbf{z}}} \leq \left(\frac{1 + \frac{1}{2q^{1/4}}}{1 - \frac{2^{r+1}}{q}} \right)^4 \cdot \left(\frac{2^{4r+10}}{q^3} - \frac{2^{3r+10}}{3q^{9/4}} + \frac{2^{r+5}}{3q^{3/4}} \right),$$

as desired.   □

Combining Lemmas 4.12, 4.13, and 4.14, we conclude that, for all $\mathbf{w} \in \ker(c_{\omega,\alpha}) \cap \mathbf{W}$ and $f_{\omega,j}(\mathbf{x}) = \gamma$, the probability that $\mathsf{HelpRec}(\mathbf{x}) = \mathsf{HelpRec}(\mathbf{x} + \mathbf{w})$ holds over choice of $\mathbf{x} \in \mathbb{Z}_q^4$ is at least

$$1 - 2\left(1 - \left(\frac{\frac{q}{2^r} - 2q^{1/4} - 1}{\frac{q}{2^r}} \right)^4 \right) - \left(\frac{1 + \frac{1}{2q^{1/4}}}{1 - \frac{2^{r+1}}{q}} \right)^4 \cdot \left(\frac{2^{4r+10}}{q^3} - \frac{2^{3r+10}}{3q^{9/4}} + \frac{2^{r+5}}{3q^{3/4}} \right).$$

Using known relationships between average min-entropy and min-entropy, we have that:

Corollary 4.15 *With all but 2^{-k} probability, the distribution over v, given the transcript of the modified protocol as well as leakage 1 mod 16 positions of $\hat{s}, \hat{s}', \hat{e}, \hat{e}', \hat{e}''$, is indistinguishable from a distribution that has min-entropy $n/8 + n/8 \cdot (-\log_2(1/2 + p)) - k$.*

4.3 Instantiating the Parameters

We instantiate the parameters as chosen in NewHope protocol: $q = 12289, n = 1024, \omega = 7, r = 2$, then we get

$$p := 2 - 2\left(\frac{\frac{q}{2^r} - 2q^{1/4} - 1}{\frac{q}{2^r}} \right)^4 + \left(\frac{1 + \frac{1}{2q^{1/4}}}{1 - \frac{2^{r+1}}{q}} \right)^4 \cdot \left(\frac{2^{4r+10}}{q^3} - \frac{2^{3r+10}}{3q^{9/4}} + \frac{2^{r+5}}{3q^{3/4}} \right)$$

$$\tag{23}$$

$$\approx 0.10092952876519123 \tag{24}$$

Therefore, under this concrete parameter setting, the distribution with leakage and transcript as defined above is indistinguishable from a distribution that has average min-entropy $128 + 128 \cdot (-\log_2(1/2 + 0.10092952876519123/2)) - 1.10910222427 \approx 237$. Moreover, with all but 2^{-80} probability, the distribution with leakage and transcript as defined above is indistinguishable from a distribution that has min-entropy 157.

4.4 On the Non-generic Part of the Analysis

Recall that in the analysis, we experimentally confirm that there exists a vector $\mathbf{w} \in \ker(\mathbf{c}_{\omega,\alpha}) \cap \mathbf{W}$.

We can support this heuristically by noting that $\mathbf{W}$ has size $(2q^{1/4})^4 = 16q$. On the other hand, the probability that a random vector in $\mathbb{Z}_q^4$ is in $\ker(\mathbf{c}_{\omega,\alpha})$ is $1/q$. So heuristically, we expect that $1/q$-fraction (approx. 16) of the vectors in $\mathbf{W}$ will also be in $\ker(\mathbf{c}_{\omega,\alpha})$.

A similar analysis can be done for other leakage patterns (n', S). Recall that our experimental attacks support the conclusion that Leaky-SRLWE is easy when the fraction of structured leakage is at least $1/4$. We may also consider parameter settings (n', S) such that $|S| = 2$ and $\frac{|S|}{n'} = 1/8$. In this case, instead of a single linear constraint $\mathbf{c}_{\omega,\alpha}$ on a single $\mathbf{x}_i$, we have two linear constraints on $\mathbf{x}_i, \mathbf{x}_{i+n/16}$. This means we will have a linear system of 8 variables and two constraints, denoted by $\mathbf{M}_{\omega,S}$. Thus, $\mathbf{W}$ will be equal to $[\frac{q}{2} \pm q^{1/4}]^8$. So the size of $\mathbf{W}$ will be $(2q^{1/4})^8 = 256q^2$ and the probability that a random vector in Z_q^4 is in $\ker(\mathbf{M}_{\omega,S})$ is $1/q^2$. So heuristically, we expect that $1/q^2$-fraction (approx. 256) of the vectors in $\mathbf{W}$ will also be in $\ker(\mathbf{M}_{\omega,S})$. Given this, the rest of the analysis proceeds nearly identically.

Acknowledgments This work was supported in part by NSF grants #CNS-1933033, #CNS-1840893, #CNS-1453045 (CAREER), by a research partnership award from Cisco and by financial assistance award 70NANB15H328 from the U.S. Department of Commerce, National Institute of Standards and Technology.

References

1. Akavia, A., Goldwasser, S., Vaikuntanathan, V.: Simultaneous hardcore bits and cryptography against memory attacks. In Reingold, O., ed.: TCC 2009. Volume 5444 of LNCS., Springer, Heidelberg (March 2009) 474–495
2. Albrecht, M.R., Deo, A., Paterson, K.G.: Cold Boot Attacks on Ring and Module LWE Keys Under the NTT. IACR Transactions on Cryptographic Hardware and Embedded Systems (2018) 173–213

3. Alkim, E., Ducas, L., Pöppelmann, T., Schwabe, P.: Post-quantum key exchange—a new hope. Cryptology ePrint Archive, Report 2015/1092 (2015) https://eprint.iacr.org/2015/1092.

4. Alkim, E., Ducas, L., Pöppelmann, T., Schwabe, P.: Post-quantum key exchange—a new hope. In Holz, T., Savage, S., eds.: USENIX Security 2016, USENIX Association (August 2016) 327–343

5. Alkim, E., Ducas, L., Pöppelmann, T., Schwabe, P.: Post-quantum key exchange—a new hope. Cryptology ePrint Archive, Report 2015/1092 (2015) https://eprint.iacr.org/2015/1092.

6. Barrow, D., Smith, P.: Spline notation applied to a volume problem. The American Mathematical Monthly **86**(1) (1979) 50–51

7. Bernstein, D.J., Chuengsatiansup, C., Lange, T., Vredendaal, C.V.: NTRU Prime . https://ntruprime.cr.yp.to/index.html

8. Blömer, J., May, A.: New partial key exposure attacks on RSA. In Boneh, D., ed.: CRYPTO 2003. Volume 2729 of LNCS., Springer, Heidelberg (August 2003) 27–43

9. Bolboceanu, M., Brakerski, Z., Perlman, R., Sharma, D.: Order-LWE and the Hardness of Ring-LWE with Entropic Secrets. In: Advances in Cryptology—ASIACRYPT 2019—25th International Conference on the Theory and Application of Cryptology and Information Security, Kobe, Japan, December 8–12, 2019, Proceedings, Part II. (2019) 91–120

10. Boneh, D., Durfee, G., Frankel, Y.: An attack on RSA given a small fraction of the private key bits. In Ohta, K., Pei, D., eds.: ASIACRYPT'98. Volume 1514 of LNCS., Springer, Heidelberg (October 1998) 25–34

11. Bos, J.W., Costello, C., Ducas, L., Mironov, I., Naehrig, M., Nikolaenko, V., Raghunathan, A., Stebila, D.: Frodo: Take off the ring! Practical, quantum-secure key exchange from LWE. In Weippl, E.R., Katzenbeisser, S., Kruegel, C., Myers, A.C., Halevi, S., eds.: ACM CCS 2016, ACM Press (October 2016) 1006–1018

12. Bos, J.W., Ducas, L., Kiltz, E., Lepoint, T., Lyubashevsky, V., Schanck, J.M., Schwabe, P., Seiler, G., Stehlé, D.: CRYSTALS—kyber: A cca-secure module-lattice-based KEM. In: 2018 IEEE European Symposium on Security and Privacy, EuroS&P 2018, London, United Kingdom, April 24-26, 2018. (2018) 353–367

13. Boyle, E., Segev, G., Wichs, D.: Fully leakage-resilient signatures. Journal of Cryptology **26**(3) (July 2013) 513–558

14. Braithwaite, M.: Experimenting with Post-Quantum Cryptography. https://security.googleblog.com/2016/07/experimenting-with-post-quantum.html Accessed: 2018-10-09.

15. Brakerski, Z., Döttling, N.: Hardness of LWE on General Entropic Distributions. Cryptology ePrint Archive, Report 2020/119 (2020) https://eprint.iacr.org/2020/119.

16. Brakerski, Z., Kalai, Y.T., Katz, J., Vaikuntanathan, V.: Overcoming the hole in the bucket: Public-key cryptography resilient to continual memory leakage. In: 51st FOCS, IEEE Computer Society Press (October 2010) 501–510

17. Chen, C., Danba, O., Hoffstein, J., Hulsing, A., Rijneveld, J., Schanck, J.M., Schwabe, P., Whyte, W., Zhang, Z.: NTRU. https://ntru.org/resources.shtml

18. Coppersmith, D.: Finding a small root of a bivariate integer equation; factoring with high bits known. In Maurer, U.M., ed.: EUROCRYPT'96. Volume 1070 of LNCS., Springer, Heidelberg (May 1996) 178–189

19. Coppersmith, D.: Finding a small root of a univariate modular equation. In Maurer, U.M., ed.: EUROCRYPT'96. Volume 1070 of LNCS., Springer, Heidelberg (May 1996) 155–165

20. Dachman-Soled, D., Gong, H., Kulkarni, M., Shahverdi, A.: On the leakage resilience of ideal-lattice based public key encryption. Cryptology ePrint Archive, Report 2017/1127 (2017) https://eprint.iacr.org/2017/1127.

21. Dachman-Soled, D., Gong, H., Kulkarni, M., Shahverdi, A.: (In)Security of Ring-LWE Under Partial Key Exposure. Journal of Mathematical Cryptology, to appear. Preliminary version in MathCrypt '19 **2018** (2018) 1068

22. Dadush, D., Regev, O.: Lecture Note of Fundamental Domains, Lattice Density, and Minkowski Theorems (2018)

23. D'Anvers, J., Karmakar, A., Roy, S.S., Vercauteren, F.: SABER. https://www.esat.kuleuven.be/cosic/pqcrypto/saber/

24. Ding, J.: A Simple Provably Secure Key Exchange Scheme Based on the Learning with Errors Problem. IACR Cryptology ePrint Archive **2012** (2012) 688
25. Dodis, Y., Goldwasser, S., Kalai, Y.T., Peikert, C., Vaikuntanathan, V.: Public-key encryption schemes with auxiliary inputs. In Micciancio, D., ed.: TCC 2010. Volume 5978 of LNCS., Springer, Heidelberg (February 2010) 361–381
26. Dodis, Y., Haralambiev, K., López-Alt, A., Wichs, D.: Cryptography against continuous memory attacks. In: 51st FOCS, IEEE Computer Society Press (October 2010) 511–520
27. Dodis, Y., Kalai, Y.T., Lovett, S.: On cryptography with auxiliary input. In Mitzenmacher, M., ed.: 41st ACM STOC, ACM Press (May/June 2009) 621–630
28. Dodis, Y., Ostrovsky, R., Reyzin, L., Smith, A.D.: Fuzzy Extractors: How to Generate Strong Keys from Biometrics and Other Noisy Data. SIAM J. Comput. **38**(1) (2008) 97–139
29. Dziembowski, S., Pietrzak, K.: Leakage-resilient cryptography. In: 49th FOCS, IEEE Computer Society Press (October 2008) 293–302
30. Ernst, M., Jochemsz, E., May, A., de Weger, B.: Partial key exposure attacks on RSA up to full size exponents. In Cramer, R., ed.: EUROCRYPT 2005. Volume 3494 of LNCS., Springer, Heidelberg (May 2005) 371–386
31. Garcia-Morchon, O., Zhang, Z., Bhattacharya, S., Rietman, R., Tolhuizen, L., Torre-Arce, J., Baan, H., O., M., Saarinen, Fluhrer, S., Laarhoven, T., Player, R., Cheon, J.H., Son, Y.: Round5. https://round5.org/
32. Goldwasser, S., Kalai, Y.T., Peikert, C., Vaikuntanathan, V.: Robustness of the Learning with Errors Assumption. In: Innovations in Computer Science—ICS 2010, Tsinghua University, Beijing, China, January 5-7, 2010. Proceedings. (2010) 230–240
33. Goldwasser, S., Kalai, Y.T., Peikert, C., Vaikuntanathan, V.: Robustness of the learning with errors assumption. In Yao, A.C.C., ed.: ICS 2010, Tsinghua University Press (January 2010) 230–240
34. Hamburg, M.: ThreeBears. https://sourceforge.net/projects/threebears/
35. Katz, J., Vaikuntanathan, V.: Signature schemes with bounded leakage resilience. In Matsui, M., ed.: ASIACRYPT 2009. Volume 5912 of LNCS., Springer, Heidelberg (December 2009) 703–720
36. Lewko, A.B., Lewko, M., Waters, B.: How to leak on key updates. In Fortnow, L., Vadhan, S.P., eds.: 43rd ACM STOC, ACM Press (June 2011) 725–734
37. Lu, X., Liu, Y., Zhang, Z., Jia, D., Xue, H., He, J., Li, B., Wang, K.: LAC: Practical Ring-LWE Based Public-Key Encryption with Byte-Level Modulus. Cryptology ePrint Archive, Report 2018/1009 (2018) https://eprint.iacr.org/2018/1009.
38. Lyubashevsky, V., Peikert, C., Regev, O.: On Ideal Lattices and Learning with Errors over Rings. J. ACM **60**(6) (November 2013) 43:1–43:35
39. Lyubashevsky, V., Peikert, C., Regev, O.: A toolkit for ring-LWE cryptography. Cryptology ePrint Archive, Report 2013/293 (2013) https://eprint.iacr.org/2013/293.
40. Malkin, T., Teranishi, I., Vahlis, Y., Yung, M.: Signatures resilient to continual leakage on memory and computation. In Ishai, Y., ed.: TCC 2011. Volume 6597 of LNCS., Springer, Heidelberg (March 2011) 89–106
41. Marichal, J.L., Mossinghoff, M.J.: Slices, slabs, and sections of the unit hypercube. arXiv preprint math/0607715 (2006)
42. Micciancio, D., Regev, O.: Worst-case to average-case reductions based on Gaussian measures. SIAM Journal on Computing **37**(1) (2007) 267–302
43. Peikert, C.: Lattice Cryptography for the Internet. In: Post-Quantum Cryptography—6th International Workshop, PQCrypto 2014, Waterloo, ON, Canada, October 1-3, 2014. Proceedings. (2014) 197–219
44. Peikert, C.: Lattice cryptography for the internet. In Mosca, M., ed.: Post-Quantum Cryptography—6th International Workshop, PQCrypto 2014, Springer, Heidelberg (October 2014) 197–219
45. Peikert, C.: How (not) to instantiate ring-LWE. In Zikas, V., De Prisco, R., eds.: SCN 16. Volume 9841 of LNCS., Springer, Heidelberg (August/September 2016) 411–430

46. Pietrzak, K.: A leakage-resilient mode of operation. In Joux, A., ed.: EUROCRYPT 2009. Volume 5479 of LNCS., Springer, Heidelberg (April 2009) 462–482
47. Regev, O.: On lattices, learning with errors, random linear codes, and cryptography. In Gabow, H.N., Fagin, R., eds.: 37th ACM STOC, ACM Press (May 2005) 84–93
48. Sarkar, S., Sengupta, S., Maitra, S.: Partial key exposure attack on RSA—improvements for limited lattice dimensions. In Gong, G., Gupta, K.C., eds.: INDOCRYPT 2010. Volume 6498 of LNCS., Springer, Heidelberg (December 2010) 2–16
49. Stange, K.E.: Algebraic aspects of solving Ring-LWE, including ring-based improvements in the Blum-Kalai-Wasserman algorithm. Cryptology ePrint Archive, Report 2019/183 (2019) https://eprint.iacr.org/2019/183.
50. Takayasu, A., Kunihiro, N.: Partial key exposure attacks on RSA: Achieving the Boneh–Durfee bound. In Joux, A., Youssef, A.M., eds.: SAC 2014. Volume 8781 of LNCS., Springer, Heidelberg (August 2014) 345–362

Correction to: The Measurement of Disaster Recovery Efficiency Using Data Envelopment Analysis: An Application to Electric Power Restoration

Priscillia Hunt and Kelly Klima

Correction to:
Chapter 3 in: M. Lee, A. Najera Chesler (eds.),
Research in Mathematics and Public Policy,
Association for Women in Mathematics Series 23,
https://doi.org/10.1007/978-3-030-58748-2_3

The original version of this chapter was inadvertently published without updating additional corrections from the author. Now, the corrections have been incorporated in chapter proof and front matter.

The updated online version of this chapter can be found at
https://doi.org/10.1007/978-3-030-58748-2_3

Printed in the United States
by Baker & Taylor Publisher Services